N

In the middle of the nineteenth century, American whaling reached its peak. Thousands of young Americans left northeastern ports annually to try their luck at catching and killing the world's leviathan. For up to four years they cruised the oceans' most desolate expanses, enduring months of tedium punctuated by days of frightening, dangerous, and dirty work.

Rites and Passages presents a social history of American whaling. Drawing on the diaries of sailors and on ship logs, this volume examines the beliefs and behaviors of men who labored at sea. It looks at the relationship between sailors and society ashore, reexamines the "tyrannical" sea captain, and studies the social dynamics of the ship's company. In particular it considers the ways in which whalemen related to women and how seafaring served as a rite of passage into manhood.

For more than a century the American public has understood whaling primarily through the work of a gifted man named Herman Melville. It is clear that other whalemen had tales to tell as well, and in *Rites and Passages* they share their compelling vision of life at sea.

Rites and Passages

In 1857, just after he had returned from a whaling voyage to the North Pacific ocean, eighteen-year-old Jamie McKenzie stood for his portrait. The resulting daguerreotype is one of the very few extant photographs of an American whaleman of the period. McKenzie, from a New Bedford family of seafarers, would sail again, but a voyage in 1862 would be his last. He was swept overboard in a storm at sea and not recovered. He was twenty-three years old. (Source: *The Bulletin from Johnny Cake Hill: A Newsletter from the Old Dartmouth Historical Society & Whaling Museum* [Fall 1987], 2.)

Rites and Passages

The Experience of American Whaling,
1830–1870

MARGARET S. CREIGHTON

Bates College

CAMBRIDGE
UNIVERSITY PRESS

Published by the Press Syndicate of the University of Cambridge
The Pitt Building, Trumpington Street, Cambridge CB2 1RP
40 West 20th Street, New York, NY 10011–4211, USA
10 Stamford Road, Oakleigh, Melbourne 3166, Australia

First published 1995

Printed in the United States of America

Library of Congress Cataloging-in-Publication Data
Creighton, Margaret S., 1949–
Rites and passages : the experience of American whaling, 1830–1870
Margaret S. Creighton.
p.   cm.
Includes index.
ISBN 0-521-43336-3 (hardback)
1. Whalers (Persons) – United States – Social life and customs.
2. Whalers (Persons) – United States – Social conditions.   3. Whalers
(Persons) – United States – History – 19th century.   4. Whaling – United
States – History – 19th century.   I. Title.
SH383.2.C74   1995
305.33'6392'0973 – dc20   94–28545
                                                            CIP

A catalog record for this book is available from the British Library.

ISBN 0-521-43336-3 hardback

Contents

Illustrations

Acknowledgments

This book seems to have taken forever, and, like so many other projects too long in the making, it carries a trail of debt. Foremost among the individuals who deserve acknowledgment are the personnel of museum libraries, special collections, and historical societies. These individuals not only suffered through unending requests for manuscripts and microfilms but also used their own long memories and historical knowledge to point me to the right places. Among the many people who gave good advice and who searched, hauled, hoisted, wheeled, or mailed, repeatedly, are Virginia Adams, Judy Downey, and Judy Lund of The Whaling Museum, New Bedford, Massachusetts; Stuart Frank and Ellen Hazen (and, long ago, Ken Martin) of the Kendall Whaling Museum, Sharon, Massachusetts; Paul Cyr of the New Bedford Free Public Library; John Koza and Kathy Flynn of the Peabody Essex Museum, Salem, Massachusetts; Nathan Lipfert of the Maine Maritime Museum, Bath, Maine; Doug Stein of the Blunt White Library at Mystic Seaport Museum, Mystic, Connecticut; Edouard Stackpole of the Nantucket Historical Association; and Philip Weimerskirch of the Providence Public Library.

A number of people have read the manuscript in parts or as a whole, years ago or more recently. James Henretta, David Hall, Cecelia Tichi, and Benjamin Labaree offered guidance and inspiration very early on. Marcus Rediker, Danny Vickers, Jeff Bolster, Valerie Burton, and Robert Schaeffer kindly read chapters or offered helpful comments in conference. Astute whaling specialists who saved me from many a nautical error (and who probably wish they could have caught more) include Bob Webb and Richard Kugler. Colin Howell and Dick Twomey, organizers of the "Jack Tar in History Conference" in Halifax in 1990, provided a remarkable forum for discussing the "new" maritime history. Kinvin Wroth generously assisted with legal research, and Susan Ransom brought skilled expertise to the job of indexing.

Lisa Norling, longtime compatriot in the social history of seafaring, deserves special mention. Over the years she has shared not only her strong intelligence, her theoretical insight, and her bibliographies but also her great gift for humor. I still laugh in memory at some of the seaport predicaments we survived.

Bates College was helpful in granting me a pre-tenure sabbatical and funds from the president's discretionary account. My colleagues and students at Bates not only reviewed parts of the manuscript, but, more important, they tolerated my exasperated conversation in hallways, through lunch, and over tea. They have my sympathy.

At Cambridge University Press, Frank Smith kept a steady eye on the course of the book, Shirley Covington saved me from a multitude of stylistic lapses, and Eric Newman knew how to crack the whip. I thank them all.

Finally, I am indebted to my parents, James Creighton and J. S. Borthwick, for continued encouragement, and to my husband, Rob Smith, for two decades' worth of steady and strong support, wise counsel, and caring generosity. Two professionals on the child-care front, Claudia Desbiens and Ann Jordan, also merit recognition. And my children, Nicholas, Malcolm, and Louisa, deserve special thanks for their forbearance. I cannot promise that on every car ride or camping trip from now on I will not be weighed down with a manuscript of one sort or another, but I *can* promise them that it will not be this one.

Archives and Collections

AAS	American Antiquarian Society, Worcester, Massachusetts
BLHU	Baker Library, Harvard University, Cambridge, Massachusetts
DCHS	Dukes County Historical Society, Martha's Vineyard, Massachusetts
EI	Essex Institute, now the Peabody Essex Museum, Salem, Massachusetts
HLHU	The Houghton Library, Harvard University, Cambridge, Massachusetts
KWM	The Kendall Whaling Museum, Sharon, Massachusetts
MeHS	Maine Historical Society, Portland, Maine
MM	The Mariners' Museum, Newport News, Virginia
MSM	Manuscript Collection, G. W. Blunt White Library, Mystic Seaport Museum, Mystic, Connecticut
NBFPL	New Bedford Free Public Library, New Bedford, Massachusetts. Courtesy of the Trustees of the Library
NHA	Nantucket Historical Association, Nantucket, Massachusetts
PC	Private Collection
PC-TB	Private Collection, Tom Bullock Jr. Courtesy of Tom Bullock, Jr., Jane Blanchard Bullock, and George Blanchard
PMM	Penobscot Marine Museum, Searsport, Maine
PMS	Peabody Museum of Salem, now the Peabody Essex Museum, Salem, Massachusetts
PPL	Nicholson Whaling Collection in the Providence Public Library, Providence, Rhode Island

Introduction:
The Passing of Nathaniel Robinson

I

Toward the end of July 1841, Reverend Daniel Lord, pastor of the Mariner's Church of Boston, returned to his office after a brief vacation. Among Lord's accumulated mail was a letter from Exeter, New Hampshire, written by Thomas S. Robinson on behalf of his twenty-one-year-old son, Nathaniel. It seems that Robinson had a favor to ask: Would it be possible for Lord to use his influence to find Nathaniel a berth on a sailing ship?[1]

Lord did not hesitate more than a day before he sent back a discouraging reply. Navigation was "dull" at present, he said, and even were it not, he knew that shipmasters did not like to take on inexperienced hands. Furthermore, he simply could not recommend seafaring. "A sailor's life," he remarked, "is a hard life." Seamen are "exposed to many hardships, [of] which those of us who reside on land, have no conception." And those hardships involved not only physical but moral dangers: "Even when our ships are commanded by a pious man, the influence of the seamen is often very bad; and many are the young men who have left home with good habits, but when they have returned, they have become disipated." Duty constrained Lord to say that "in nine cases out of ten it would be much better for [a young man] to find some employment on shore."[2]

The year following in the Robinson household is a year largely lost to the historical record. We do know, though, that Nathaniel Robinson was not dissuaded from pursuing a sea voyage. In fact, according to a family member, his anxious determination to join a ship's crew had obtained "full possession of

[1] Daniel M. Lord to Thomas S. Robinson, 30 July 1841, PMS.　[2] Ibid.

his mind." And in the late spring of 1842, Robinson's father sent Daniel Lord a second request for assistance. This time Lord delivered an even more emphatic warning: "And as to my own advise, I would say to your son *be content to seek employment on the land. . . .* Should he go I am persuaded he will regret it."[3]

Lord's repeated admonitions fell on Nathaniel Robinson's deaf ears. One year from the day that Lord sent this last letter, the New Hampshire boy was in New London, Connecticut, aboard the whaleship *Nantasket* and bound for the Pacific Ocean.

Nathaniel Robinson, like many other new sailors, made a journal of his voyage, and we read it in suspense, anxious to see whether he vindicated his own or the wary minister's expectations. The diary, along with a letter he sent home, reveals that he did both. Writing to his sister Mary on the fourth of July 1843 (after twelve days at sea), Robinson stressed the rightness of his decision. His officers, he assured her, were "all professors of religion," and the ship was nothing less than "an old Methodist meeting house." His shipmates were "very civil & well disposed," and the one black seaman in his watch was "the most good natured negro I ever . . . saw & full of fun." In addition, Robinson was as comfortable as he could be. The ship was so dry that "not a bit of spray comes over the side," and he was perfectly satisfied with forecastle fare. Just to prove that he had plenty to eat every day, he mentioned relishing a "huge slice" of gingerbread given to him in celebration of the national holiday.[4]

A few shadows, however, extend over Robinson's letter. He admitted to his sister that "it is with difficulty I can conform myself to my situation." "The habits of the Sailors," he told her, "the dirty forecastle, foul air &c are averse to my feelings. . . . I know I should not like it, though you need not say anything about it." In his private diary, Robinson was even more critical of his surroundings. Far from being bountiful and palatable, the food was actually fit for nothing better than "a swill pail." The salt horse he found "disgusting," and the molasses tasted like "tar." The worst affront to his sensibilities, though, was a pan of food containing a "huge swine's snout with the bristles all on!" In public the forecastle had been a comfortable abode, but in private he condemned it as a "hole."[5]

Daniel Lord had warned Nathaniel Robinson's father that seafaring transformed young men and that the experience dissipated and hardened them. And the story of Nathaniel Robinson seems to underscore the minister's opinions. How, we wonder, would this young man of "good habits" fare in a world that challenged his Victorian tastes and threatened his well-being? Would he, like

---

[3] *The Exeter News-Letter*, 23 October 1843, p. 3, col. 3; Daniel M. Lord to Thomas S. Robinson, 15 June 1842, PMS.

[4] Nathaniel G. Robinson to Mary F. Robinson, 4 July 1843, PMS.

[5] Diary of Nathaniel G. Robinson, *Nantasket*, 26, 28, 29 June 1843, PMS.

the sailors described in Lord's cautionary letter, return home socially and morally unrecognizable?

This young forecastle hand cannot answer these questions for us. On July 14, during his night watch, when he was sitting on some spars on deck, Robinson was "seized by a slight hacking cough." Walking only made the cough worse, and he soon discharged "nearly a gill of blood." The sight of the sick sailor alarmed the shipmaster, who, having left the Azores the previous day, tacked ship and returned to port. As he was lowered into a boat to be rowed to shore, Robinson heard the mate remark, "Poor fellow, I pitty him."[6]

Ashore in a Faial hospital, Nathaniel Robinson managed to keep up his diary. With a shaky hand and an increasingly weaker hold on his pencil, he recorded his slow demise. He hoped that the "Lord will spare my life to reach home once more." But then, as if the burden of the days' events were too much, he simply listed the dates: "Thursday Sep 1st Friday 2d Sat 3rd Sun 4." He had enough stamina to note that the consul's brig, which would attempt to return him to Boston, would sail on Wednesday, September 13. At the bottom of his last entry, made on that vessel, stood a message to anyone who might assist him: "15 of the white drops in a little Water if I bleede."[7]

Nathaniel Robinson returned home still alive. With the help of a fellow seaman, he survived the trip back to Boston, and on the evening of October 7, 1843, he arrived at the entrance to his family's house. "His return," reported the local newspaper, "was wholly unexpected. When his father met the carriage at the door, he could scarcely believe his eyes that it was really his son, so great had been the ravages of disease." Twenty-four hours later the young sailor was dead.[8]

In a sense, this discussion of nineteenth-century whaling takes up where Nathaniel Robinson's sea voyage left off. Robinson's untimely death leaves us with questions that lie at the heart of the seafaring experience. What happened to land-bred men when they went sailing, sometimes for years at a time, sometimes for life? What of home did they take with them to sea, and what of home did they relinquish? What did they learn from a life in far latitudes, in the confines of small wooden ships, among men of many cultures?

Over the past century, these questions have been posed, discussed, and even heatedly debated. Numerous historians and contemporary observers in the late nineteenth and early twentieth centuries readily dismissed seamen as odd and even alien species, implying that saltwater either attracted peculiar types or genetically transformed men. "I have much to do with these people," commented an American consul in 1874, "and find them to be a distinct and peculiar species

---

[6] Ibid., 14 July, 1843.   [7] Ibid., 22 August to 16 September 1843.
[8] *The Exeter News-Letter*, 23 October 1843, p. 3, col. 3.

of the genus homo – a kind of floating gypsy." "[The sailor] lives in a little world of his own," echoed a historian in 1900. For "the majority of his days he lives in utmost ignorance of what is going on in the world. He is like the inhabitants of some undiscovered country. . . ." Another historian, writing in 1933, claimed that a sailor was a man "set apart in habit, thought, and also by a certain aptitude born of the sea. . . . He was a rigorous example of the molding and seasoning of men to an end – the service of ships and the sea."[9]

More recent discussions of North American seamen have brought mariners back within the human fold. Jesse Lemisch was one of the first historians to argue that the image of the socially alienated sailor demands critical scrutiny. In 1968, in his highly influential essay, "Jack Tar in the Streets," Lemisch dismantled the notion of mariner as misfit, claiming that such a stereotype was an enduring fabrication of the ruling class. Jack Tar, he suggested, had been usefully denigrated as an outsider because in reality he challenged the political, economic, and cultural interests of people in power.[10]

Canadian scholars have also been instrumental in reclaiming seamen from the ranks of outcasts. David Alexander, in particular, furthered the assertion that mariners were not an anomalous population but, rather, "working men who got wet." Other important Canadian work has emphasized the degree to which ships and shipboard cultures were (and are) products of evolving human societies ashore and of landward economic development.[11]

But others insist that in our zeal for humanizing mariners we should not underestimate the transformative or unique aspects of the sailing experience or dismiss the ways that seamen "expressed . . . opposition to shoreside norms." For centuries, one scholar maintained recently, Atlantic seamen were "shaped by shipboard cultural forms that endowed them with distinctive attitudes and perceptions." Historians of eighteenth-century Anglo-American seamen have made a strong case for the degree to which sailors' struggles at sea, against both human and natural adversaries, engendered a special willingness to challenge

[9] J. Grey Jewell, *Among Our Sailors* (New York: Harper, 1874), p. 14; Frank T. Bullen, *The Men of the Merchant Service* (New York: Stokes, 1900), pp. 251–3; Felix Riesenberg, *The Log of the Sea* (New York: Harcourt, Brace, 1933), p. 28. Among other accounts that stress the distinctive nature of whalemen are Clifford Ashley, *The Yankee Whaler* (Garden City: Halcyon, 1942), p. xxvii; and Alexander Starbuck, *History of the American Whale Fishery* (1877; reprint, Secaucus, N.J.: Castle Books, 1989), p. 1. For an additional survey of such perspectives see Eric Sager, *Seafaring Labour: The Merchant Marine of Atlantic Canada, 1820–1914* (Kingston: McGill–Queen's University Press, 1989), p. 3.

[10] Jesse Lemisch, "Jack Tar in the Streets: Merchant Seamen in the Politics of Revolutionary America," *William and Mary Quarterly* 25 (July 1968): 371.

[11] See the essays by Alexander and ten other contributors in Rosemary Ommer and Gerald Panting, eds., *Working Men Who Got Wet* (St. John's: Maritime History Group, 1980); also Sager, *Seafaring Labour*, p. 3.

authority. Translated into action in port, this radicalism gave sailors leading roles in revolutionary activity around the world.[12]

This study of a group of American whalemen at sea in the mid-1800s extends the discussion about the nature and effects of seafaring and argues that the experiences of whalemen were inescapably shaped by land and sea both. The ships these Americans sailed and the waters they traversed were the setting of social rites and ocean passages marking both the tenacious hold of the American homeland and the transformations of landsmen into sailors.

In the first instance, American whaling cannot be considered apart from the varied landward interests that sustained it. The technology of the ship, the contours of seafaring work, the structure of social relations at sea all reflected the cultural and economic dynamics of the mainland.

On a personal level as well sailors linked themselves to the land they left behind. They saw the whaling voyage as a "return" trip, and even from the farthest reaches of the globe they cast their minds homeward. Whalemen before the mast may have been particularly land-oriented because they were usually "greenhands" (novices fresh from shore), but even "old salts" reveal the strong pull of their home base.

Whalemen not only actively sustained links between land and sea; they carried aboard the multiplicity of American society, especially the differing perspectives of race, age, and economic station. They also brought to sea with them their home-grown interests as men, as gendered individuals. Maritime historians have rarely studied or even noted this aspect of a mariner's identity and have generally assumed that most sailors embraced a single, unchanging masculine style – usually tough and antidomestic. But recent insights into the ways that masculinity and womanhood are constantly shifting and being redefined challenge us to consider the ways that gender shaped the sailor's experience.[13] Close examination of seafaring records, for instance, reveals how much

---

[12] The quotations are taken from W. Jeffrey Bolster's essay, " 'To Feel like a Man': Black Seamen in the Northern States, 1800–1860," *Journal of American History* 76 (March 1990): 1178. On sailors' special radicalism, see Lemisch, "Jack Tar" and the important work of Marcus Rediker, especially *Between the Devil and the Deep Blue Sea: Merchant Seamen, Pirates, and the Anglo-American Maritime World, 1700–1750* (Cambridge, UK: Cambridge University Press, 1987); also Peter Linebaugh and Marcus Rediker, "The Many-Headed Hydra: Sailors, Slaves and the Atlantic Working Class in the Eighteenth Century," *Journal of Historical Sociology* 3 (September 1990): 225–51. On the distinctiveness of seamen and their influence ashore in the twentieth century, see Bruce Nelson, *Workers on the Waterfront: Seamen, Longshoremen, and Unionism in the 1930s* (Urbana: University of Illinois Press, 1988).

[13] Essays describing the wide range of (changing) shoreside masculine ideals can be found in Mark C. Carnes and Clyde Griffen, eds., *Meanings for Manhood: Constructions of Masculinity in Victorian America* (Chicago: University of Chicago Press, 1990); see especially the essay by Nancy F. Cott, "On Men's History and Women's History," in this volume. See also Ava Baron, ed., *Work Engendered: Toward a New History of American Labor* (Ithaca:

some sailors struggled to adhere to one masculine code and how others balked at being cast in a uniform mold. Mariners set sail with no single formula about how to be a better "man."

If whaling sustained the social continuum between sea and shore, it also had the capacity to change men. This work contends that even though sailors held fast to land-bred identities and social styles and sailed in circumstances dictated ashore, they were not unaffected by the sailing experience. The work whalemen performed, the sea world they encountered, the relationships they sustained, and the conflicts they endured all left their impressions and took a toll.[14] Deepwater experience sometimes reinforced mens' given allegiances and their long-held beliefs. But it also nurtured new social bonds and fostered new frames of mind. The sea could be a destabilizing element in more ways than one.

II

American whaling in the middle decades of the nineteenth century represented seafaring at its extreme. Voyages lasted up to four years, and ships spent much of their time not docked and loading but cruising on some of the oceans' most desolate expanses. During this period, the American whaling fleet reached its height in terms of its size, and according to many historians, it was the heyday of the fishery. It was at this time that whaling became the third biggest business in Massachusetts, and the American fleet the largest in the world. It was at this time that whalers made important exploratory forays into the Pacific and toward the two poles, and when the international influence of whalemen was so felt (and so feared) that reformers and missionaries staged global efforts both to counter the "immoral" effects of these men and to convert sailors to their own causes.[15]

Cornell University Press, 1991); E. Anthony Rotundo, *American Manhood: Transformations in Masculinity from the Revolution to the Modern Era* (New York: Basic Books, 1993); Joan W. Scott, "Gender: A Useful Category of Historical Analysis," *American Historical Review* 91 (1986): 1053–75. Recent work analyzing the experience of British and American seafaring through the lens of gender appears in Margaret Creighton and Lisa Norling, eds., *Iron Men and Wooden Women: Gender and Maritime History, 1700–1900* (Baltimore: Johns Hopkins University Press, forthcoming).

[14] This argument is similar to one that Daniel Vickers put forth in a review essay several years ago. Mariners, he said, "were situated between two poles." They were pulled on the one hand by "hometown culture and landward expectations" and on the other by the "generally collective and radicalizing force of their work experience." "Roundtable," *International Journal of Maritime History* 1 (December 1989): 312–13.

[15] Elmo P. Hohman, *The American Whaleman* (New York: Longmans, Green, 1928), p. 7; Herman Melville, *Moby Dick* (New York, 1851; reprint, New York: Norton, 1967), pp. 99–100; Starbuck, *American Whale Fishery*, pp. 98–9; David Moment, "The Business of Whaling in the 1850's," *Business History Review* 31 (Autumn 1957): 264; *The Fourth Annual*

Nineteenth-century whaling has had many chroniclers – the number of published reminiscences on the subject would alone fill a nautical library – but only a few individuals have studied the industry critically. Elmo Paul Hohman is without a doubt the best-known authority on the social and economic history of the fishery, and his 1928 account of whaling labor and capital in the 1800s remains the definitive work on the topic. Hohman stressed the hardships of whaling and attributed much of it to owners' greed and to the destructive imperatives of the profit motive. But whalemen's misery also reflected, he argued, the "irresponsibility, vice, depravity, and criminality" of crews themselves. Although this work does not share Hohman's sort of moralizing, it nevertheless owes a great deal to his monumental study. It is also indebted to recent work on the colonial fishery, to new assessments of whaling communities ashore, and to a statistical survey of the industry's economics.[16]

The account that follows is based to some extent on evidence generated on land – newspaper accounts, crew lists, legal evidence, and correspondence. But it is drawn chiefly from materials produced on shipboard, particularly the diary and logbook testimony of nearly 200 sailors. These informants were white Americans, most of whom hailed from coastal New England. Like the majority of whalers of this period, these men were deepwater seamen, that is, men who worked in the Indian, Pacific, Arctic, or southern Atlantic oceans.[17] They represent all levels of the ship's hierarchy – from captain to cabin boy – but most are either seamen or greenhands who lived before the mast or men who inhabited the aftercabin as shipmasters and mates.[18] The focus of the work is undeniably "fore" and "aft" – on the geographic, social, and political polarities of the whaleship.

Report of the American Seaman's Friend Society (New York, 1832), p. 11; Hugh Davis, "The American Seaman's Friend Society and the American Sailor, 1828–1838," *American Neptune* 34 (January 1979): 45.

[16] Hohman, *American Whaleman*, p. 59. Other important analyses include Daniel F. Vickers, "Nantucket Whalemen in the Deep-Sea Fishery: The Changing Anatomy of an Early American Labor Force," *Journal of American History* 72 (September 1985): 295; Edward Byers, *The Nation of Nantucket: Society and Politics in an Early American Commercial Center 1660–1820* (Boston: Northeastern University Press, 1987); Edouard A. Stackpole, *The Sea Hunters: The New England Whalemen During Two Centuries 1635–1835* (Philadelphia: Lippincott, 1953); Lisa Norling, "Contrary Dependencies: Whaling Agents and Whalemen's Families 1830–1870," *Log of Mystic Seaport* 42 (Spring 1990): 3–12; Lance E. Davis, Robert E. Gallman, and Teresa D. Hutchins, "Productivity in American Whaling: The New Bedford Fleet in the Nineteenth Century," *NBER Working Paper Series* 2477 (December 1987); idem., "Risk Sharing, Crew Quality, Labor Shares and Wages in the Nineteenth Century American Whaling Industry," *NBER Working Paper Series* 13 (May 1990).

[17] Teresa Dunn Hutchins, "The American Whale Fishery 1815–1900: An Economic Analysis" (Ph.D. diss., University of North Carolina, 1988), p. 78; Starbuck, *American Whale Fishery*, pp. 422–532.   [18] See Appendix 1 for further information on these diarists.

Examined collectively, journals and logbooks illuminate life and work under sail with an immediacy that other sources cannot begin to approximate. As valuable to seafaring research (albeit with as many limitations) as slave narratives are to the study of the plantation, students of seafaring would do well to spend more time with these materials. That scholars have not tapped this sort of resource to any large extent may reflect the paucity of social historians who tackle maritime subjects or it may indicate that researchers have been unaware that such accounts exist.[19] One scholar claimed in the 1920s that there "was a dearth of authentic records of American whaling from the pens of participants. . . ." Another, writing thirty years later, lamented the fact that "unfortunately not many whalemen kept diaries or journals." More recently, a historian commented that many ships' logs are "silent" or "blank," making it difficult to understand seafaring from the inside. "We can no longer enter the sailors' communities," he noted, "or eat their food or see with their eyes. . . ." "The ordinary sailor," a fourth historian remarked, ". . . is hard to find in the historical record, but emerges instead, however unrealistically, in the images of Dana, Melville, Conrad and O'Neill."[20]

This study, which draws so heavily on sailors' journals, necessarily disputes many of these assertions. Yet the foregoing comments do contain important truths. There is no question that the interior lives of many seamen, particularly nonliterate seamen, are not reflected in these historical records. And even though the authors of these diaries represent diverse backgrounds and ages, they were all white and all American.[21] Sailors were, as one mariner put it, a "mixed and motley company," and seafaring as a whole was an enterprise involving many races and nationalities. A study of American whaling in its totality would take into account the perspectives of men of color, men of different ethnicities, and men from many countries.[22]

Such a study would consider, for instance, the perspectives of African American seamen, who, although diminishing in numbers on merchant ships

[19] Stuart C. Sherman called attention to the importance of whaling logbooks and diaries some time ago. See his *The Voice of the Whaleman* (Providence: Providence Public Library, 1965). A study that makes excellent use of large numbers of seamen's diaries to evaluate whalemen's literary expression is Pamela Miller's *And the Whale is Ours: Creative Writing of American Whalemen* (Boston: David R. Godine and the Kendall Whaling Museum, 1979). A detailed annotated diary is *The Captain's Best Mate: The Journal of Mary Chipman Lawrence on the Whaler* Addison *1856–1860*, ed. Stanton Garner (Providence: Brown University Press, 1966).

[20] Clifford Ashley, *The Yankee Whaler* (New York: Halcyon House, 1926), p. xxv.; Stackpole, *Sea Hunters*, p. 396; Sager, *Seafaring Labour*, p. 222. Marion Diamond, "Queequeg's Crewmates: Pacific Islanders in the European Shipping Industry," *International Journal of Maritime History* 1 (December 1989): 140.

[21] See Appendix I for diarists' ages, where known.

[22] The quotation is taken from the diary of Henry Davis, unknown vessel, 14 May 1862, MSM; A recent history of world whaling that emphasizes the international dimensions of the industry is Richard Ellis, *Men and Whales* (New York: Knopf, 1991).

after 1840, still found a place of employment on whalers.[23] It would incorporate the viewpoint of sailors of Cape Verde and the Azores. These Portuguese islanders were so important to the fishery that by the post–Civil War period, according to Briton C. Busch, one-fourth of the crews of whaling and sealing vessels were from these islands. Shipowners, hoping to cut costs in challenging market settings, actively recruited Atlantic island labor, and natives of these islands accepted low pay and the rigors of whaling work as an alternative to ekeing out a living in economically depressed homelands. Whaling also offered a means of passage to the United States.[24]

A study of European, Latin American, and particularly Pacific island sailors would also contribute to a more complete history of American whaling. Captains faced with deserting American sailors in Pacific ports signed on Polynesians, particularly Hawaiian natives, for cruises in the Pacific and the Arctic. They viewed these islanders – whom they saw as "skilled and daring sailors" and who were, like Portuguese seamen, paid poorly relative to American seamen – as preferable to "less reliable" American hands. Eventually, says historian Marion Diamond, whaling captains "made a practice of sailing with a skeleton crew, to which [they] added extra sailors recruited closer to the killing grounds."[25]

The sailors whose remarks inform this account, then, represent a domestic facet of an international labor force. This is not to say that these sorts of men were a minority aboard whaleships. Crew lists and whaling contracts tell us that white, literate, northeastern Americans were actually typical of many of the men who entered the fishery in the middle 1800s, who sailed from American ports, and who returned in American vessels.[26] Ships' crews were most culturally diverse during the midyears of a voyage, as vessels lost, discharged, and recruited men in foreign ports.[27]

Although these men were typical to some degree of outgoing and incoming

[23] Bolster, "To Feel like a Man," p. 1194; Mary Malloy and Stuart M. Frank, "'Heroes in the Ships': African Americans in the Whaling Industry," *Catalogue and Object List*. Sharon, MA: The Kendall Whaling Museum, 1990. James Farr claims that black men served on "some New Bedford whaleships in considerable numbers." See his "A Slow Boat to Nowhere: The Multi-Racial Crews of the American Whaling Industry," *Journal of Negro History* LXVIII (Spring 1983): 164–5.

[24] Briton Cooper Busch, "Cape Verdeans in the American Whaling and Sealing Industry, 1850–1900," *American Neptune* 45 (1985): 104–16; Malloy and Frank, "'Heroes in the Ships.'"   [25] Diamond, "Queequeg's Crewmates," pp. 128–9, 140.

[26] A recent study of more than 36,000 New Bedford shipping contracts signed in the mid-1800s indicates that in the early 1840s over 80 percent of whaling crews signed their names to agreements and that on the eve of the Civil War, almost 75 percent could do so. See Lance E. Davis, Robert E. Gallman, Teresa D. Hutchins, "Risk Sharing, Crew Quality, Labor Shares and Wages in the Nineteenth Century American Whaling Industry," *NBER Working Paper Series* 13 (May 1990): 27. Crew lists describing the nativity and citizenship of whalemen are examined here in Appendixes III and IV. (And see cautionary note.)

[27] See Appendix IV.

American crews, it is likely that there were more privileged men among them than sailed in an average American vessel. Among these informants are a number of men – approximately one-fifth of the diarists – who hailed from the same region and were the same age as their fellow greenhands but who viewed themselves as outsiders to seafaring labor and life. These men were the Richard Henry Danas of the ship's company who, by virtue of their "talent and information"[28] (or, as we might put it, education and wealth), dissociated themselves from other sailors.

Robert Weir was one such individual. Having signed aboard the *Clara Bell* in 1856, he regarded his compatriots in the whaling service as "loose and worthless." He claimed that "their very touch [was] contaminating – their slightest presence pollution." It gave him the "blues" to think that his "refinements & politeness [were] fast scattering to the four winds of heaven." Another of these outsiders claimed that the "habits of sailors [are] averse to my feelings." A third denounced his shipmates as "the most wicked set of men that I ever encountered."[29] These men, proud advocates of Victorian culture, frequently moralized on aspects of self-control: temperance, punctuality, sexual continence, and thrift.

The many whalemen who hailed from laboring backgrounds ashore – from farms, factories, or seafaring communities – sometimes embraced the same social ideals as their outsider counterparts, but frequently they did not. They also did not alienate themselves from the whaling community, even if they found the deep sea experience distasteful.[30] The dynamic social and intellectual tensions between "outsiders" (who are also referred to as "Victorian seamen") and their shipmates is discussed in much of the account that follows.

Whalemen had various reasons for keeping journals and for taking notes in logbooks. First mates (and sometimes shipmasters) maintained a logbook as a

[28] The phrase comes from the Diary of Elias W. Trotter, *Illinois*, 17 April 1846, Log 1005 A & B, WMLOD.

[29] Diary of Robert Weir, *Clara Bell*, 9 May 1856, MSM; Nathaniel G. Robinson to Mary F. Robinson, 4 July 1843, PMS; Diary of Charles Stedman, *Mt. Wollaston*, 11 January 1854, NBFPL.

[30] The distinction frequently made in this work between Victorian outsiders and other whalemen reflects an important social tension on shipboard, but it does not at the same time imply that these were the only cultural divisions among seamen. As historians of labor culture on land have informed us, working people established diverse cultural milieux. See Bruce Laurie, *Working People of Philadelphia, 1800–1850* (Philadelphia: Temple University Press, 1980) and Baron, ed., *Work Engendered*. For earlier important analyses of cultural tensions among workers in antebellum America, see Herbert G. Gutman, "Work, Culture, and Society in Industrializing America," *American Historical Review* 78 (June 1973): 531–88; Paul Faler, "Cultural Aspects of the Industrial Revolution," *Labor History* 15 (Summer 1974): 367–94; Daniel Walker Howe, "American Victorianism as a Culture," *American Quarterly* 27 (December 1975): 507–32; Paul Johnson, *A Shopkeeper's Millennium: Society and Revivals in Rochester, New York 1815–1837* (New York: Hill & Wang, 1978).

matter of professional responsibility. Maritime law required that the mate, as the first officer was called, keep a detailed account of the events of a voyage, including information on wind direction, vessel course, and noon position. Whaling logkeepers were sometimes quite lax about recording position, but they did make concerted attempts to note whales taken and seen, to remark on the weather, and to report unusual shipboard events. Such reports provided owners and other masters with valuable details about whaling grounds and whale migration patterns, as well as important information on poorly charted foreign seas. In addition, they provided important testimony about any shipboard matter that might come under later dispute. As author and lawyer Richard Henry Dana remarked, the ship's log was "a very important trust." It was, he said, the "depository of the evidence of everything that may occur during the voyage; and the position of the ship, the sail she was under, the wind, &c., at any one moment, may become matters of great consequence to all concerned."[31]

The fact that logbooks were "evidence" was not lost on their keepers, and there were occasional disputes in the aftercabin over whose history of a voyage officials would eventually see. Clothier Peirce, mate of the bark *Rodman*, diligently reported in the *Rodman*'s official log that the captain had struck him "a Blow with his fist" and that during a skirmish the captain had pointed a revolver at his head. "Brave man," he wrote. Shortly after this last entry, though, the handwriting in the log changes dramatically. "This day," the captain noted, "put the Mate off duty for disobedience of orders, for gross insult to me and for cruelty to crew." The shipmaster then began to enter his own testimony in the dispute, such as the allegation that the mate had slandered him to the sailors. "For instance," noted the captain, "he told the Steward that he had rather have Capt Smith's Old Shoes to conduct a voyage than to have my whole body." For this and for other sorts of disrespect, the master not only disrated the officer and separated him from the log but discharged him five months later.[32]

The captain and the mate of the *Polar Star* settled their disagreements over ship evidence by staging a cutting fight over logbook pages. According to the second mate, who quietly observed the goings on, the shipmaster came below on June 11, 1860, and quite casually asked the first officer "How the Log was getting along." The mate replied that he had three logs. When asked to produce the "Log Book that belonged to the Ship," the mate announced that his past entries would be offered "to no one but the owners when the ship arrived at New Bedford." He then drew his knife to cut out certain pages and "mutilated"

[31] Sherman, *Voice of the Whaleman*, pp. 30–31, says whalemen were "quite casual" about recording the ship's position; Richard Henry Dana, *The Seaman's Friend* (Boston: T. Groom, 1851), p. 145.
[32] Diary of Clothier Peirce and Elisha Babcock, *Rodman*, 10–11 September 1857; 4 February 1858, Log 548, WMLOD.

the book. He was then ordered to his stateroom. The second mate, perhaps eyeing a more advanced berth and perhaps hoping that the captain would request to see *his* version of things, noted carefully in his account that he could not understand how anyone could be disrespectful or insulting toward "Capt Weeks" as he was a "Gentleman in evry respect."[33]

Although official logbooks help to ground the study here, whalemen's personal diaries provide the evidential backbone of this work. Some of these diaries were epistolary accounts or records intended for a family member or a sweetheart at home. Lewis Williams on the *Gratitude* wrote his account for "Mother and the rest of the folks," and he enjoyed noting occurrences because "it seems to carry me nearer home and I believe I am nearer home for doing it." Samuel Chase also used his journal to lessen the distance between him and his home. With it, he felt, he could soar above the ocean's vast expanses and "be with them in thought if not in person." Ambrose Bates wrote his journal for his wife, Annie. Mindful of the risks involved in whaling, though, he also left it for posterity: "If it be my lot to leave this form of clay upon some corral bed. Or upon the desolate shore of a frozen zone. I shall not go unmourned for I believe there are those who would miss me many a long day. And perhaps this book may reach them and these unworthy museings of mine may cheer their remembrance of me."[34]

Seamen also kept journals for themselves. Men who sought a raise in rank or who wanted to be seamanlike made careful notes of wind, whale sightings, and sail changes. George Dyre, a first-time sailor, was handed a new book as he departed from home, along with directions for its use. "George," wrote his uncle, "you are now about to embark on a whaling voyage and enter upon a new career of life – perhaps a few words of advice from an experienced person may be of benefit to you." George was reminded to note down everything he could, including the force of the wind, the passing of "large flocks of birds, Gulfweed, kelp," and the latitude and longitude.[35]

Journals served not only the interests of the upwardly mobile mariner but also the young man eager to improve his position in life more generally. Journal keeping was a means, said nineteen-year-old James Allen, "to improve my writing." Samuel Chase, aboard the *Arab* in 1842, similarly claimed that making daily entries "gives a person practice in composition [and] penmanship and I am shure it is time well (at least not ill) employed."[36]

Greenhands, who were wide-eyed at the newness of life at sea, were among the most enthusiastic and prolific keepers of journals. They took to their pens

[33] Diary of John S. P——— (name illegible), *Polar Star*, 11 June 1860, NBFPL.
[34] Diary of Lewis Williams, *Gratitude*, 11 June 1859, PMS; Diary of Samuel Chase, *Arab*, 9 October 1842, MeHS; Diary of Ambrose Bates, *Milwood*, 2 April 1867, KWM.
[35] Diary of George N. Dyre, *Napoleon*, December 1858, Log 835, WMLOD.
[36] Diary of James Allen, *Alfred Gibbs*, 25 July 1870, PPL; Chase diary, 2 November 1842.

Joseph Ray, whaleman aboard the ship *Edward Cary*, decorated the frontispiece of his journal with this elaborate watercolor sketch. Such designs are not as rare as one might expect, and many whaling libraries boast an illustrated journal or a nicely calligraphed logbook. Why did whalemen decorate their diaries? The easiest answer is that they had plenty of time.

to record everything from the first bout of seasickness to the first smashed whaleboat, and they described shipboard life and work, landfalls, and liberty days with observant detail. Veteran seamen authored diaries, too, in their case not to record novelty but to relieve dull familiarity. Third mate Edwin Pulver was among many experienced whalemen who turned to his journals when boredom got the better of him. At the end of his voyage aboard the *Columbus* in 1852, he penned the following poem in honor of diary keeping:

> Fairwell old journal I love you well
> Because of by gone days you tell
> And I love you for other reasons too
> One is because you allways gave me something to do.[37]

Whether they kept diaries for the first time at sea or perhaps the last, whalemen inevitably focused on scenes of drama and conflict: struggles with whales, battles with storms, and confrontations between officers and crews. Seafaring journals, therefore, highlighted the unusual, the exciting, and sometimes the unhappy circumstance. Seaman Samuel Chase noted that "indeed, it often is, if not generaly the case, that when we have the greatest variety of subject to enter in our journals there is something unpleasant to note, some disagreeable event or accident, which we would have been better pleased had it not transpired." Journal keepers wrote when they had the time to write, of course – at the end of the day or during a dry spell of whaling. When they were busy chasing or trying out whales, they were frequently too tired or, one might be tempted to say, too content to bother. James Allen, aboard the *Alfred Gibbs* in 1870, suggested that such was the case with him. He wanted to "give it up" with respect to journal keeping, largely because there was little to write about: "Here I am trying it again tonight my subject is this a cloudy day, in a whaleship, everything going on nicely. . . ."[38]

In every logbook collection in historical archives there are scores of terse, repetitive logs that do not bear much resemblance to the journals used in this study. These, perhaps, are accounts of whaling success, or of voyages on which social peace and quiet satisfaction prevailed. It is important to bear in mind, then, that behind the accounts of the writers here may stand another whole range of experience that, for better or for worse, was devoid of "interest."

This study focuses on whalemen at sea in the mid-1800s, but it moves its lens away from shore slowly. The first chapter examines the history of the whaling industry, and particularly the history of the landward society that moved men seaward, from the colonial period until the late nineteenth century. The second chapter centers on the period from 1830 to 1870 and looks at how whaling crews assembled on shore, first from the perspective of owners and then through the eyes of would-be seamen themselves. Chapter 3 shifts the focus to where it will remain: on men at sea. This chapter describes whaling work and considers

---

[37] Diary of Edwin Pulver, *Columbus*, 15 October 1852, PPL.
[38] Chase diary, 20 November 1842; Allen diary, 29 September 1870.

the ways that this work affected sailors' social relations and gender identity. Chapters 4 and 5 explore the experience of seafaring from opposite ends of the ship's hierarchy. The investigation begins with the shipmaster in the aftercabin and moves to a study of sailors living in the forecastle. Chapter 6 follows whalemen into foreign ports. It describes sailors' social life on leave and examines the degree to which mariners' shipboard relations extended to sojourns ashore. Chapter 7 discusses whalemen and women, sexual identity, and shipboard intimacy. The Afterword speculates on the legacy of the deepwater experience for those who left the sea and questions the extent to which sailors who went whaling for life were true social renegades.

This study relies heavily on evidence found in sailor diaries and ship logbooks. The spelling, punctuation, and capitalization in these sources have been retained in most cases. When meaning seemed seriously compromised, slight adjustments have been made in punctuation, and clarifying words have been inserted in brackets. The generic term *whalemen* is used throughout this book. In all cases it refers to the writers here, or to those individuals they describe.

1

The Evolution of the American Whale Fishery, 1650–1900

I

During the peak of American whaling, in the middle of the nineteenth century, nearly 700 whaleships, carrying 20,000 men, might be found at sea in a given year.[1] No one knew for sure what was in store for the men who set sail from American shores, for they faced the vagaries of the ocean environment. But the outcomes of their voyages involved human decision as much as meteorologic luck because the configuration of sailing ships and sailing crews reflected the enterprise and determination of men and women ashore. During the two centuries that saw the rise and decline of the whale fishery, the individuals who set these ships in motion skillfully adjusted to evolving economic needs, to developments in technology, and to the changing circumstances of animal ecology. How they shaped the industry is the subject of this chapter.

The American whale fishery as a systematic enterprise had its beginnings in the seventeenth century. Colonists, who brought with them from Europe an appreciation of the whale's value and who needed trade commodities, began to hunt the animal from the Long Island shore and in Buzzards Bay sometime in the mid-1600s. Setting up stations on the beach, they posted lookouts who surveyed the ocean for spouting or breaching. At the first sign of whales, they dispatched six-man crews to harpoon the animals and tow the carcasses back to shore.[2] The

---

[1] Elmo P. Hohman, *The American Whaleman: A Study of Life and Labor in the Whaling Industry* (New York: Longmans, Green, 1928), p. 26.

[2] Harry D. Sleight, *The Whale Fishery on Long Island* (Bridgehampton: Hampton Press, 1931), p. 4; Daniel F. Vickers, "Nantucket Whalemen in the Deep-Sea Fishery: The Changing

object of these hunts was primarily the right whale, a nontoothed whale whose blubber when boiled down produced a crude oil useful for lights and lubrication. These cetaceans, which average forty-four feet in length, are to be found in (among other places) the North Atlantic. From cold, northern waters where they feed, they migrate south along the eastern edge of the continent to winter and breed in the relatively warm waters between Cape Cod and the Carolinas.[3]

Noting that mainland residents were trading successfully in whale oil, and well aware that they were close to right whales' path of migration, Nantucket Island settlers began to eye the seas with the idea of entering the business. Recognizing the advantages of whaling was not the same as knowing how to prosecute the hunt, however, so in the mid-seventeenth century Nantucketers invited an experienced Long Island whaleman to come to live with them. Despite offers of subsidy and housing, he never arrived. Finally, in 1690 a Cape Cod native named Ichabod Paddock accepted an invitation to live on Nantucket and work with the islanders to launch the fishery there. The instruction was obviously a success, for by the 1700s Nantucket was doing steady business in right whale oil with Boston and had the largest whaling fleet in the colonies.[4]

Whalemen continued to hunt from shore through the eighteenth century, but they also began to venture farther afield. Legend has it that an intrepid Nantucketer named Christopher Hussey was responsible for extending the frontiers of the fishery. In 1712 Hussey was hunting for right whales when he was blown off course into deep water. There he came face to face with a sperm whale, killed it, and brought its oil-filled body back home. The toothed sperm

Anatomy of an Early American Labor Force," *Journal of American History* 72 (September 1985): 281; Hohman, *American Whaleman*, pp. 41–2. The question of who exactly the first American whalemen were has been debated. "According to some writers," noted Alexander Starbuck, the first American whalemen were Indians. These men took advantage of whales that stranded along the seacoast and also pursued the animals in canoes. Richard Kugler claims that the idea of Native American whalemen systematically pursuing whales is a mistaken "popular belief." Although white colonists hired Indians to row whaleboats, he says that Indians "brought to whaling no traditions or prior experience of their own." There is little evidence, Kugler states, beyond one observer of Penobscot Indians in 1605 to "sustain the assertion of an independent whaling tradition by native Americans." See Alexander Starbuck, *History of the American Whale Fishery* (1877; reprint, Secaucus, NJ: Castle Books, 1989), p. 5; and Richard C. Kugler, "The Whale Oil Trade, 1750–1775," in *Seafaring in Colonial Massachusetts* (Boston: Colonial Society of Massachusetts, 1980), p. 156.

[3] Edward Byers, *The Nation of Nantucket: Society and Politics in an Early American Commercial Center, 1660–1820* (Boston: Northeastern University Press, 1987), p. 79; Daniel Vickers, "The First Whalemen of Nantucket," *William and Mary Quarterly* XL (October 1982): 562. An excellent survey of whale species and whale behavior can be found in Richard Ellis, *Men and Whales* (New York: Knopf, 1991), pp. 1–32. For whales that frequent New England coastal waters, see also Steven Katona, David Richardson, and Robin Hazard, *A Field Guide to the Whales & Seals of the Gulf of Maine* (Rockland, ME: Maine Coast Printers, 1975), p. 2; pp. 23–5; Hohman, *American Whaleman*, p. 181.

[4] Starbuck, *American Whale Fishery*, p. 17.

whale, which has a reservoir of pure, light, liquid oil in its head, may have been known (and sought) since the late seventeenth century, but Hussey's discovery set off the first full-scale hunt for this wide-ranging mammal.[5]

The search for sperm whales was spurred by several additional developments. A burgeoning colonial population created a large market for whale products, and British demand for whale oil expanded. London, according to one historian, was the "best-lit city in the world," and its numerous street lamps were fueled by whale oil. In addition, England's proliferating textile mills needed whale oil for cleansing wool, and its construction industry used it as a paint base. Demand for sperm oil in particular increased with improvements in processing. In the mid-1700s, an enterprising individual figured out how to separate the head-matter of the sperm whale into two products: oil and a solid, waxy substance known as spermaceti. Spermaceti made exceptional candles. Advertised in Boston in 1748 as "exceeding all others for Beauty, Sweetness of Scent when extinguished . . . [and] emitting a soft, easy, expanding light," these candles became one of the most marketable of whale products. In addition, sperm oil itself was recognized as an illuminant superior to right whale oil, and from then on it was increasingly distinguished and marketed apart from its cruder counterpart.[6]

American merchants and shipbuilders facilitated the hunt for sperm whales with two technological innovations. First, they modified their long, square-sterned whaleboats into agile, speedy, double-ended craft. Whalemen were therefore better able to pursue the fast-swimming sperm whale and better equipped to maneuver around the teeth and flukes of this mammal when things turned close and nasty.[7] Second, they added tryworks to their whaleships. Tryworks were large iron pots built into brickwork on a ship's deck, and they allowed whalemen to "try out," or boil down, whale blubber at sea. Whereas sailors had always interrupted a hunt to carry blubber to shore for rendering, these traveling processors permitted a more continuous and more distant whalehunt. Whalemen thus became, in the words of one scholar, "truly pelagic hunters."[8]

Encouraged by changing technology and high market prices, northeastern merchants sent whaleships fanning out onto the world's oceans. Decade by

[5] Ibid., p. 15.

[6] Byers, *Nation of Nantucket*, p. 141; advertisement in Kugler, "Whale Oil Trade," p. 155; sperm oil market information in Vickers, "First Whalemen," p. 565.

[7] Kugler, "Whale Oil Trade," p. 157; Vickers, "First Whalemen," p. 565.

[8] John R. Bockstoce, *Whales, Ice, and Men: The History of Whaling in the Western Arctic* (Seattle: University of Washington Press and New Bedford Whaling Museum, 1986), p. 27; Kugler, "Whale Oil Trade," p. 154.

(*Opposite*) A topsail schooner maneuvers in the bay and three men prepare to launch a rowing boat in this early-nineteenth-century view of the town of Sherburne on Nantucket Island.

This drawing, photographed from a wall exhibit at the New Bedford Whaling Museum, illustrates the five whales most commonly hunted in the nineteenth century. At the top is the legendary **sperm whale**, made infamous in Herman Melville's *Moby Dick*. It is a toothed whale, prized for its reservoirs of clear oil. It inhabits all except the most frigid of the world's waters. Second from the top is the **right whale**. The right whale is a baleen whale, so called for the flexible yet bony "food strainers" that hang from the roof of its mouth. The right whale was hunted both for its oil and its baleen and was the whale of choice (hence "right") for American colonists. Right whales made relatively easy targets for sea-bound colonists, because they migrated to and from the Arctic along the North American coast. The middle whale is the Arctic **bowhead whale**. The discovery of

decade they discovered new "grounds" – areas where cetaceans fed, mated, or bred.[9] By the late 1790s, American whalemen were hunting sperm and right whales in the South Atlantic and had entered the Pacific. By the end of the first decade of the nineteenth century, they were searching the waters of the southern Indian Ocean, the higher latitudes of the South Pacific, and all along the western rim of South America. In at least one way, the industry reached its apogee in this period: Whale oil exports constituted almost 53 percent of all pounds sterling earned in New England from exports to Great Britain. Whaling was thus more important to the region's economy than it had ever been or would later become.[10]

The American whale fishery not only changed in geographic and technologic scope over the course of the 1700s; the social dimensions of the voyage modified as well. A man who served in the fishery in the early industry integrated his life as a whaleman with his life as a landsman to a degree unknown to later sailors. In the early eighteenth century, he left shore for only a few hours or perhaps a day at a time, regularly returning home to his family at night. He hunted whales seasonally and dovetailed his seafaring work with his work on land. Later in the century, as ships took to deep water for longer periods of time and voyages increased to nearly five months, whaling life and home life became more separate.[11]

As the industry became a bigger and more expensive business throughout the 1700s, the social structure on shipboard also changed. Owners, who spent more

[9] Hohman describes thirty grounds in use by whalemen by 1840. Fifteen were in areas of the Pacific. *American Whaleman*, pp. 148–50.

[10] John G. B. Hutchins, *The American Maritime Industries and Public Policy, 1789–1914* (New York: Russell & Russell, 1941), p. 270; Hohman, *American Whaleman*, p. 38; Bockstoce, *Whales, Ice, and Men*, pp. 27–8; Byers, *Nation of Nantucket*, p. 144.

[11] Daniel Vickers, "Maritime Labor in Colonial Massachusetts: A Case Study of the Essex County Cod Fishery and the Whaling Industry of Nantucket, 1630–1775" (Ph.D. diss., Princeton University, 1981), pp. 269–72.

new bowhead populations in the western Arctic in the mid-1800s helped extend the life of the American whaling industry. The slow-swimming bowhead is thick with blubber, but more important is that it has plentiful, long baleen. The whale second from the bottom is the **humpback whale**. This whale is the darling of late-twentieth-century spectators, migrating near shore and being distinctly acrobatic. To nineteenth-century whalemen, though, it was for several reasons less sought after than the whales described above: its oil was less marketable, its baleen was short, and the humpback tended to sink after it died. The bottom whale in the drawing, the **California gray whale**, was the object of a short, intensive hunt in the mid–nineteenth century in its breeding grounds in the lagoons of Baja California. Because it lacked baleen of commercial importance, it was not the object of more widespread hunting. Source: wall text, New Bedford Whaling Museum.

and more money to build and maintain steadily larger ships, began to distinguish the men who navigated the ships from the rest of the ship's crew. They recruited captains who were specifically trained and experienced and who could be relied on to protect owners' investments and interests abroad.[12]

To cement their relationships with captains, owners offered them selectively higher reimbursement. In contrast to the practice in the early shore fishery, where owners compensated whalemen equally, no matter whether they steered the boat or manned the oars, merchants now instituted a pay scale. They gave masters a share of the proceeds, called a *lay*, which was around 1/16th (one out of every sixteen barrels, in other words). They also encouraged masters to invest in their own voyages if they wished to. Ordinary whalemen, by contrast, were given lays ranging from 1/32 and 1/36th and were rarely, if ever, owners. Merchants furthered the inequalities of the take by increasing their own share. In the early fishery they had taken the proceeds of one-quarter of the bone and oil and left the remaining three-quarters for the crew to divide equally. Now, in the late eighteenth century, they began to take a larger percentage for themselves.[13]

There seems to have been little mobility between ranks and between labor and ownership in whaling throughout the 1700s. On Nantucket, according to Daniel Vickers, the positions of ships' officers were the white islanders' "exclusive preserve," and owners were almost always English colonists. On the other hand, the arduous jobs of pulling whaleboats went to those "most limited in individual liberty." During the days of the shore fishery, and into the days of deep-sea whaling, this meant men who were nonwhite, mostly Native Americans. On Nantucket, Wampanoags handled whaleboats almost exclusively. The key to their employment, says Daniel Vickers, was "coercion." Drawn into indebtedness, Indians turned over their labor and their earnings to merchant creditors. They rarely (if ever) became shipowners themselves, and colonists, suspicious of the Indians' independent ideas and their "irregular" work habits, felt they had good reason to keep it that way.[14]

The whaling merchants of Nantucket did not always look to the Wampanoag tribe to work their whaleboats. At the end of the 1700s, the growing fishery needed more and more workers, but the population of the tribe was declining. Shipowners turned inland to find men to hire. Although they continued to seek out racial minorities, they also began to recruit another disadvantaged group: propertyless sons of white colonists. Attracted to the fishery by the potential for quick fortune, these men would have been able to avoid whaling a generation earlier. But now, many of them lacked purchasing power and some were in debt

[12] Vickers, "Maritime Labor," pp. 156–7, 171, 274–82; Kugler, "Whale Oil Trade," p. 156; Hohman, *American Whaleman*, p. 28.   [13] Byers, *Nation of Nantucket*, p. 93.

[14] Vickers, "Maritime Labor," pp. 170–82; 305; Starbuck, *American Whale Fishery*, pp. 11–26; Edouard A. Stackpole, *The Sea Hunters: The New England Whalemen During Two Centuries 1635–1835* (Philadelphia: Lippincott, 1953), pp. 16–17.

to merchants, and they were forced to do the sort of seafaring work that "their grandfathers would have rejected out of hand."[15]

II

The business and practice of whaling during the industry's golden age of the mid-1800s paralleled in many ways the developments of the colonial period. Merchants invested more and more money in larger vessels, which sailed for longer and more distant voyages. They took more and more of a share of a vessel's proceeds, gave more to officers, and left less for unskilled hands. They heightened the division of labor within the ship. They also kept a sharp eye out for changing markets and diminishing returns, and, thus, despite rapidly changing economic circumstances, they stayed their profitable course.

The period from the American Revolution until after the War of 1812 was a time of great volatility for the fishery. After the Revolution, British markets for oil closed and French markets opened. Then French markets closed. The whaling fleet suffered serious damage and loss during wars and embargoes, and British competition took its toll on the business.[16] Following the political crises, economic depressions, and other disruptions of the early national era, however, whaling entered a period of fast expansion. The discovery of new sperm and right whale grounds complemented growing markets. On the domestic front, whale oil was in increasing demand as an illuminant in developing commercial centers and as a lubricant in proliferating factories. Sperm oil became the preferred means of lighting lighthouse lamps in the early 1800s, and, for those who could afford to acknowledge the difference, spermaceti candles rather than "foul-smelling" tallow candles became the illuminant of choice.[17] Markets in Britain opened again, too, as that country's whaling enterprise diminished and when in 1843 Great Britain lowered the tariff on imported whale oil.[18]

The sperm whale fishery, long the sustenance and pride of Nantucket, did not remain centered there. The island was disadvantaged by a shallow harbor, distance from markets, and a diminishing supply of labor. By the mid-1800s, most whaling merchants had moved their operations to the mainland. Although Nantucket continued to be an active presence in the industry – it sponsored seventy-five whalers in 1846 – nearby New Bedford soon rightly claimed to be the fishery's focal point. In 1846 it registered 254 whaling vessels, and in 1857 it controlled half of the nation's whaling business in terms of management, property, and processing. Other ports were involved – New London, Sag

---

[15] Vickers, "Nantucket Whalemen," pp. 295–6.

[16] Teresa D. Hutchins, "The American Whale Fishery 1815–1900: An Economic Analysis" (Ph.D. diss., University of North Carolina, 1988), pp. 22–38.

[17] Hohman, *American Whaleman*, p. 39; Byers, *Nation of Nantucket*, p. 249.

[18] Teresa D. Hutchins, "American Whale Fishery," pp. 48–56.

This nineteenth-century lithograph illustrates the wide range of uses for whale products, including baleen for umbrella stays, whale oil for manufacturing, sperm oil for illumination, and ambergris, a whale's intestinal secretion, valuable as a binding agent for perfumes. The aboriginal interest in the whale as food was not shared by American people in general.

Harbor, Fairhaven, and Nantucket especially – but New Bedford far and away dominated the industry.[19]

The story of whaling's success in the mid–nineteenth century is a story of entrepreneurial readjustment. Industry owners made a major shift in the geography of the fleets in the mid-1800s in response to significant changes in the domestic market for whale products. What happened, quite simply, was that

[19] Hohman, *American Whaleman*, pp. 41–2. The ports in southern New England and Long Island Sound were most active in the fishery, but other towns from Maine to New Jersey sent whaleships and whalemen to sea in the mid-1800s. See Kenneth R. Martin, *Whalemen and Whaleships of Maine* (Brunswick, ME: Harpswell Press, 1975), and Ellis, *Men and Whales*, pp. 160–1.

This cartoon from *Vanity Fair* takes the whales' perspective in celebrating the discovery of oil wells in Pennsylvania in 1859. Reflecting the political and social crisis of the impending Civil War, it also expresses Unionist sentiment.

Americans found new sources of light. Spurred in part by the high prices of sperm and right whale oil, developers had been looking for more accessible (and hence cheaper) illuminants for some time. The discovery of oil wells in Pennsylvania in 1859 settled their search.[20] Coal oil (kerosene) could not compete with sperm oil as a lubricant of fine machinery or with right whale oil as a crude lubricant for heavy machinery, but it could outdo all whale oils as a source of cheap and effective light. One observer wryly described whale oil street lamps in New York City as being so low in light "that it seemed as if they were mourning for the loss of the moon." Wall Street whale lamps, likewise, were so dim that they could not even compete with "five and twenty full grown lightning bugs."[21] Little wonder that urban residents, among other citizens, preferred the higher intensity of kerosene.

The marketing of new illuminants jarred the optimism of whaling merchants, but two developments helped to keep the whaling business alive. The first was a rising demand for whalebone (baleen), and the second was the discovery of new populations of whales that were loaded with this material.

Baleen consists of horny slats, fringed with hair, that extend in rows from the

[20] Starbuck, *American Whale Fishery*, p. 110.
[21] Teresa D. Hutchins, "American Whale Fishery," p. 259, quoting Frederick L. Collins, *Consolidated Gas Company of New York: A History* (New York, 1934), p. 57.

upper palate of nontoothed whales. These slats help the whale strain its food:
the animal takes in large quantities of sea water filled with crustaceans and
plankton and then squeezes the water out through this filter. Baleen had a
number of commercial uses. When it was cut into strips, heated, and then
shaped, it became a tough yet flexible material marketable for umbrellas, whips,
and walking canes. It was put to greatest advantage, though, in the fashion
industry. Whalebone provided structural support for corsets and helped to
round out voluminous Victorian skirts.[22]

Right whales yielded baleen, of course, but it was the Arctic bowhead that
became the whaleman's prize. The bowhead, which sometimes yielded over 300
barrels of oil and 3,000 pounds of whalebone (compared with 130 barrels and
1,000 pounds for the right whale), helped extend the life of the whaling industry.[23]

Finding bowhead whales in the western Arctic Ocean was, according to John
Bockstoce, "the most important whaling discovery of the nineteenth century."
It is credited to an enterprising shipmaster named Thomas Roys, who in July
1848 sailed through narrow Bering Strait into the Arctic Ocean. The western
Arctic had been off-limits to whalemen because of the sharp ice that regularly
clogged its waters, and shipmasters saw little reason to endanger their vessels by
entering it. Roys, however, changed matters. On July 25 he and his crew took
a strange whale – a "new fangled monster" as his crew called it – and to everyone's delight it turned out to be thick with blubber and loaded with baleen. It
was a bowhead whale.[24] For the next fifty years, a good proportion of the
American whaling fleet risked Arctic ice and centered its hopes on this animal.

The discovery of bowhead whales in Arctic waters was one of several factors
behind the shift of American whaling to the Pacific in the mid-1800s. The
observation that whale species were being depleted closer to home and the
establishment of Hawaii and Panama as centers for the transshipment of whale
products made it both necessary and feasible for whalers to base themselves in
that ocean. Voyages necessarily extended in length as a result and not infrequently
lasted four years. If owners had had their way, voyages might have been much
longer, for merchants were painfully aware of the time and money that they lost
in sending ships from the northeastern United States around one of the two
capes to their eventual cruising grounds.[25]

[22] Charles M. Scammon, *The Marine Mammals of the North-western Coast of North America*
(1874; reprint, New York: Dover, 1968), p. 17; David Moment, "The Business of Whaling
in America in the 1850's," *Business History Review* 31 (Autumn 1957): 265; Hohman,
American Whaleman, p. 148; Bockstoce, *Whales, Ice, and Men*, pp. 29, 208.
[23] Hohman, *American Whaleman*, p. 148; Bockstoce, *Whales, Ice, and Men*, p. 95; Moment,
"Business of Whaling," p. 265.   [24] Bockstoce, *Whales, Ice, and Men*, pp. 23–4.
[25] Lance E. Davis, Robert Gallman, and Teresa D. Hutchins, "Productivity in American
Whaling: The New Bedford Fleet in the Nineteenth Century," *NBER Working Paper Series*
2477 (December 1987): 8–9. On the often conflicting desires of owners, captains, and
insurers on voyage length, see the correspondence of whaling agent Jonathan Bourne Jr.,

BARK   SHIP

Largely because it was more maneuverable, the bark rig was increasingly favored over the ship rig for deepwater whaling.

The whaleships that carried men around the world's circumference in the middle of the nineteenth century were significantly larger than the schooners and brigs used in the earlier fishery.[26] The average size of a midcentury whaler built for global work was around 100 to 150 feet long and 25 feet wide, and ranged from 250 to 400 tons.[27] Owners favored two types of three-masted vessels for deepwater whaling at this time: the square-rigged ship,[28] which took a crew of at least thirty men, and the bark, which often accommodated a slightly smaller crew. On barks, the mizzen (stern-most) mast was rigged fore and aft.

1856–58, Mss 18, S-G, Ss 2, vols. 1–3, WMLOD, and the correspondence of the agency of Swift & Allen, 1858–59, Mss 5, S-G 3, S-F, Ss 1 and 2, WMLOD.

[26] Vessels engaged in the Atlantic fishery in the 1800s tended to be much smaller than those operating in other areas. According to Erik A. R. Ronnberg, Jr., schooners in Atlantic whaling from 1845 to 1860 averaged 110 to 120 tons and brigs, 140 to 148 tons. These vessels composed from 2 to 10 percent of the total whaling fleet in the mid-nineteenth century. See Ronnberg, "Whaling Brigs and Schooners: A Preliminary Survey." Lecture given to Peabody Museum Associates, November 1970, WMLOD (with thanks to Mr. Ronnberg).

[27] Richard Ellis, *Whales and Men* (New York: Knopf, 1991), pp. 181–3. Understanding a vessel's tonnage, in all its mathematical complexity, is a daunting task. Simply put, tonnage describes a ship's cargo-carrying capacity and is determined by measuring volume. The term came from *tun*, and referred originally to the number of tuns of wine a ship could carry. For a discussion of the formulas used to calculate tonnage throughout the nineteenth century and a history of register tonnage, see John Lyman, "Register Tonnage and Its Measurement," *American Neptune* 5 (1945): 223–34.

[28] The term *ship* is used throughout this account but does not necessarily refer to this rig. It more generally signifies a sailing vessel.

Such a rig was handled more easily than the ship rig and was favored by men who negotiated the tricky, ice-filled waters of the Arctic.[29]

Whaling merchants designed vessels to make the most of the changing hunt and to facilitate whaling work. They also shaped the physical space aboard ship to accommodate the division of labor they established and to reflect their sense of social propriety. A detailed look at the allotment of responsibility and work aboard a whaler, as well as the designation of space aboard ship, shows how this floating institution worked.

Whaling crews of the mid-1800s were highly specialized. They included a shipmaster, three or four mates, three to five boatsteerers, a cook, a steward, and a cabin boy, at least one artisan, such as a cooper or blacksmith, and around fifteen skilled and unskilled seamen.[30]

Shipowners gave the captain a wide range of power and responsibility, and for this they awarded him a lay of around 1/15th – a share about twelve times larger than that of a greenhand.[31] The captain was expected to have skills not only as a navigator, a sailor, a boatman, and a whaleman but also to be a financial agent, a business manager and last, but not least, a supervisor of personnel. Ashore in foreign ports he managed the hiring and firing of sailors and engineered the grim process of hunting, pursuing, and punishing deserters. At sea he walked the delicate line between pleasing his owners and pleasing his sailors enough to prevent mutiny or massive desertion. He expected obedience from his officers and at the same time relied on them to be as strict with the crew as he was with them. In the day-to-day running of the ship, he might stand lookout himself or even head a whaleboat. Or he might remain on deck for the duration. He sometimes dealt directly with the men, either with solicitous attentions or with coercive brutality, or he let the mate do the honors.

The shipmaster lived along with the rest of the ship's officers in the after part (the stern) of the vessel, but he was granted more privacy and more room than any other man aboard ship. His quarters included a daycabin (perhaps 8 feet by 10 feet) with lockers and usually a built-in parlor sofa. His cabin led into a private toilet and sink area and also into his stateroom. In this area, which might

[29] Davis et al., "Productivity in American Whaling," p. 10; Table 7. The size of a whaleship's crew varied depending upon whether the vessel carried four or five whaleboats. A five-boat whaleship needed a crew of thirty just to man the boats.

[30] Hohman, *American Whaleman*, p. 119; Lance E. Davis, Robert E. Gallman, and Teresa D. Hutchins, "Risk Sharing, Crew Quality, Labor Shares and Wages in the Nineteenth Century American Whaling Industry," *NBER Working Paper Series* 13 (May 1990): 4–33. These three economists conclude that as the nineteenth century progressed, the size of crews increased about 9 percent, an increase that was accompanied by changes in numbers at various ranks. There was an expansion in the number of greenhands, officers, and boatsteerers and a reduction in the number of skilled and semiskilled seamen.

[31] This average is based on a study of New Bedford voyages destined for the Pacific whaling grounds from 1840 to 1858 and 1866. The study is presented in Davis et al., "Risk Sharing," pp. 51–6 (Table 2). All lay data described here derive from this source.

be 10 feet long, he maintained a clothes closet and, if he prized a level night's sleep, a gimballed bed. Otherwise his berth was built in.[32]

The chief, or first, mate was the master's right-hand man. The position of mate was, according to a whaleman who had just assumed this position, a "hard task." In the first place, "there is a large crew to attend to and keep busy some of them know something and some does not. So from the 2nd mate to the cook are depending upon me to know this and that and what they shall do next." The mate was called on not only to direct work but also to keep track of the ship's position and to keep the voyage record. "I have the lat to look after as shure as the day comes," reported this same officer, "then to get an altitude in the afternoon and then when night comes and I have been busy all day and feel like turning [in] I have to set down and write first I begin with my journal then comes the ships log a long days work to write up."[33] For his troubles the mate received an average "short" lay of around 1/24th.

The duties of the lesser mates were manifold. Along with the first mate, they worked in shifts to supervise the sailing of the ship and the work of the crew on deck. During the whale hunt itself, all officers headed boats, directed the chase, and wielded the lance to kill the whale.[34] These men, who earned shares of the catch ranging from 1/40th to 1/70th, lived and dined aft, within the cabin complex. The mate and perhaps the second mate had private staterooms probably measuring no more than 6 feet by 4 feet. The lesser mates shared quarters. All the officers, or the "afterguard" as they were sometimes collectively known, ate in shifts at a table in the center cabin.[35]

Stationed in the ship's hierarchy between officers and foremast hands and housed, appropriately enough, amidships in a separate compartment called the "steerage" were men who performed a wide range of work. These included the boatsteerers, who were skilled whalemen charged with harpooning whales and steering whaleboats during the hunt, and who headed watches on cruising grounds. These men, who often hailed from the Azores or the Cape Verde Islands, had to have a proven ability to dart an iron and handle a steering oar. They usually commanded lays slightly under 1/90th.

The steerage gang, as sailors sometimes termed them, included not only boatsteerers but also a number of artisans. Coopers were hired – for around the 1/60th lay – to help build casks and to keep them in good order and sometimes

---

[32] Reginald B. Hegarty, *Birth of a Whaleship* (New Bedford: New Bedford Free Public Library, 1964), p. 78; Anonymous pamphlet, "The Whaleship Charles W. Morgan" (Mystic, CT: Mystic Seaport Museum, 1991).

[33] Diary of Ambrose Bates, *Nimrod*, 2 December 1860, KWM.

[34] According to Charles Scammon, if the captain headed a boat, then the fourth mate became his boatsteerer. See *Marine Mammals*, p. 222.

[35] Hohman, *American Whaleman*, p. 119; p. 126; Francis Allyn Olmsted, *Incidents of a Whaling Voyage* (1841; reprint, Rutland, VT: Charles E. Tuttle, 1970), pp. 51–2; Hegarty, *Birth of a Whaleship*, pp. 78–9; Anonymous, "The Whaleship Charles W. Morgan."

to tend the ship when the whaleboats were down. Blacksmiths, who worked on irons and lances and innumerable other forged items, and all-purpose carpenters might also be part of the steerage group. These men earned considerably less than coopers, around 1/170th. Other steerage residents included the cook and steward, who on the ships surveyed here were predominantly men of color, and sometimes a cabin boy. The "boy," who earned the longest lay of any individual on the vessel, averaging around 1/400th, was assigned any number of tasks, some related to sailoring and some connected with cabin work. The steward was in many ways the captain's private servant. He kept the aft section of the ship, as one steward put it, in "apple pie order." The man who was associated with the steward in rank and reimbursement – the cook – had a very difficult if not to say impossible job. Without wasting provisions but frequently drawing on rotten or rancid material, this individual was expected to provide the aftercabin with palatable food and the forecastle with ample nourishment. More often than not, his was a thankless set of tasks.[36] He and the steward earned around the 1/140th lay.

And then there were the men before the mast. These foremast hands included unskilled seamen (greenhands), who earned an average lay of 1/189th, and semiskilled (ordinary) and skilled (able) seamen, who received lays ranging from 150th to 170th. These men were assigned to everything from greasing masts to cutting potatoes, to rowing captains from ship to shore, to tarring rigging, to scraping paint, to cleaning. Especially to cleaning. Early every day they swept and scrubbed decks – "servant's maid's business," some grumblingly called it. Most important, though, these men were charged with steering the ship, handling sail, looking for whales, and catching them. They may have been expected to acquire some skills above their station, so that they could move into a boatsteerer's job if they had to, but they were not regularly instructed in officers' work such as navigation. It was in the interests of masters and mates to keep some of the tools and tricks of the trade to themselves.

Foremast hands were housed together in the legendary forecastle, a space under the main deck in the bow of the vessel, perhaps 16 to 25 feet in length, with little headroom. This was home to about fifteen to twenty men and also accommodated sea chests and built-in berths. These berths, each about 6 feet by 2 feet, were set in double rows, one above the other. Prisms placed in the upper deck (the forecastle's ceiling) offered some light to the area, but the main source of sun and air was the forecastle companionway, the single stairway that led up to the deck.[37]

[36] Hohman, *American Whaleman*, p. 120; Davis et al., "Risk Sharing," p. 4; Diary of Silliman B. Ives (under the pseudonym Murphy McGuire), *Sunbeam*, 18 June, 16 July 1871, Log 618, WMLOD.

[37] For a description of the physical layout of the forecastle, see Hohman, *American Whaleman*, p. 126, and Hegarty, *Birth of a Whaleship*, p. 79–80. Hegarty gives the dimensions of a typical forecastle as 18 by 20 feet.

The spatial layout of the American whaling ship reflected the special interests of the men who financed the industry and the more general ideals of business society. In contrast to eighteenth-century whalers – which, by virtue of their size if nothing else, were relatively intimate social spaces – these ships honored prevailing cultural divisions with physical boundaries. Men who were disadvantaged by their color, their inexperience and income, or the fact that they performed "women's" work were generally housed together forward or in the steerage.[38] The officers of whaling vessels, who were predominantly white men of middle-class interests, usually lived by themselves aft. Owners not only partitioned the shipmaster off from "inferiors" but also granted him sufficient space aboard the ship to be able to honor that critical marker of Victorian status and respectability: the separation between home and work. The captain could retreat, if he wanted, to the private, home-like space of his parlor or private stateroom, where he sometimes even sequestered his wife, or he could venture forth to the very public, very male space of the ship at large.

The boundaries drawn on these ships are consistent with those favored by the emerging urban middle class in northeastern America. Numerous historians have described a changing social landscape in cities and towns where employers and wage earners were increasingly separated at work and at home, where lines between races were tightened, and where the spheres of the sexes were ideally becoming more distinct.[39] The whaleship as it evolved also bore a growing resemblance to that quintessential institution of the industrial revolution: the factory. In the concentration of unskilled men before the mast, in the separation between knowledge and power aft and muscles and brawn forward, in the specialized but interdependent functions of men, and in the management of labor by an overseer, the whaleship of the mid-1800s was considerably more factory-like than were its predecessors.

We can go only so far in claiming that the nineteenth-century whaleship was a modern institution, however. As much as whaling owners may have sought efficient, well-regulated operations at sea, whaling was fully tied to the vagaries of nature. The whale hunt was sporadic and unpredictable, and work was highly task-oriented. Sailors with a dead whale labored nonstop and with backbreaking intensity to process the animal, but when their hunt was unsuccessful, their work rhythms slowed to a snail's pace. Furthermore, even though officers and

---

[38] Hohman describes how Portuguese whalemen performed washing for other sailors of the forecastle. *American Whaleman*, p. 54. On the gender division of labor at sea, see also Chapter 6, this book.

[39] Paul E. Johnson, *A Shopkeeper's Millennium: Society and Revivals in Rochester, New York, 1815–1837* (New York: Hill & Wang, 1978), pp. 52–5. Johnson's book is one among many good accounts of the passing of the household economy and of the consequent changes in social ideals and social structures. For a discussion of how spatial arrangements reflect social status in general and gender hierarchies in particular see the important work by Daphne Spain, *Gendered Spaces* (Chapel Hill: University of North Carolina Press, 1992).

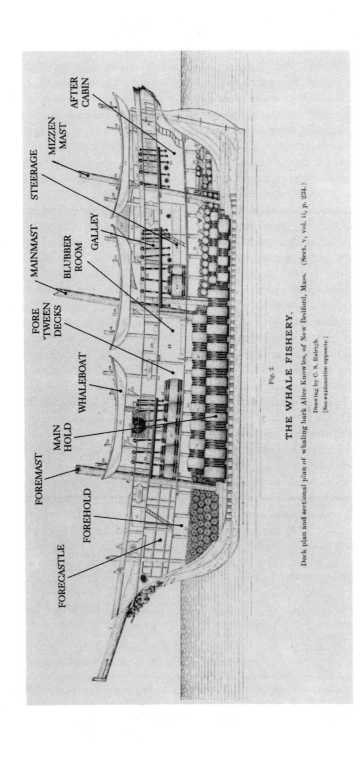

FORECASTLE

FOREMAST

MAIN
HOLD

FOREHOLD

WHALEBOAT

FORE
'TWEEN
DECKS

MAINMAST

BLUBBER
ROOM

GALLEY

STEERAGE

MIZZEN
MAST

AFTER
CABIN

Fig. 2.

THE WHALE FISHERY.

Deck plan and sectional plan of whaling bark Alice Knowles, of New Bedford, Mass.   (Sect. v, vol. ii, p. 234.)

Drawing by C. S. Raleigh.

[See explanation opposite.]

crew members lived separately, had distinct work assignments, and earned unequal levels of reimbursement, they were linked to one another by the lay system, and they worked in traditional ways, face to face. They collaborated in the killing of whales and in the process of cutting in and trying out. During the hunt itself, in fact, officers and boatsteerers and foremast hands each had representatives in the six-man boats and had to work together in tight coordination. Their lives depended on it.[40]

Whaleships themselves cost an impressive amount of money to build and equip. Figures vary from year to year and from one rig or size to the next, but both contemporary analysts and historians suggest that a whaleship plus outfit might have cost from $40,000 to $50,000 during the decade before the Civil War.[41] Owners safeguarded their investments in a number of ways. Like their predecessors in the colonial fishery, they diminished personal risk by sharing ownership. They divided up their property between a group of partners on a fractional basis – usually in multiples of eight. An individual could own an eighth or a sixty-fourth of a ship, for instance. Owners often had family ties or at least business connections with one another. A co-owner might be a ship chandler, a baker, a sailmaker, or a cousin or a brother. Shipmasters sometimes owned a share of the vessel as well.[42]

The operation of the whaling business – purchasing vessels, managing recruitment, outfitting the ships, supervising the voyage, handling reimbursements, and selling the catch – rested with a managing owner, also known as the ship's agent. The agent usually held the largest share in the vessel, commonly 4/32nds, and was further reimbursed for his additional work. In the port of

---

[40] Discussions about the degree to which sailing ships in modernizing Anglo-America can be understood as protofactories can be found in "Roundtable," and Marcus Rediker, "The Common Seaman in the Histories of Capitalism and the Working Class," both in the *International Journal of Maritime History* 1 (December 1989): 311–40. A classic essay on how work habits and labor discipline changed with the advent of industrial capitalism is E. P. Thompson, "Time, Work-Discipline and Industrial Capitalism," *Past and Present* 38 (1967): 56–97.

[41] Teresa D. Hutchins, "American Whale Fishery," p. 175; Hohman, *American Whaleman*, p. 325; Martin Joseph Butler, "J. and W. R. Wing of New Bedford: A Study of the Impact of a Declining Industry Upon An American Whaling Agency," (Ph.D. diss., Pennsylvania State University, 1973), p. 45; Jonathan Bourne, Jr., to Captain F. A. Weld, 11 November 1859, "Jonathan Bourne, Jr., Business Records," Mss 18, S-G, Ss 2, vols. 1–3, WMLOD.

[42] Butler, "J. and W. R. Wing of New Bedford," p. 23. For further discussions of the shipmaster as owner, see Chapter 4.

(*Opposite*) A cross-section of the bark *Alice Knowles* gives a good indication of the arrangement of space aboard a whaleship. Note the distance between the forecastle, the living space for newly hired and less-skilled seamen, and the aftercabin, where the officers resided. The steerage, where lesser officers and ships' artisans lived, is also visible.

Visitors to the New Bedford waterfront around 1870 pose before an impressive array of whaling technology: row upon row of casks, and the bark *Massachusetts*, drying sail.

New Bedford alone in the 1850s, over 100 agents competed for their share of the world's whales. The largest owners – with names like Jonathan Bourne, Charles Tucker, Isaac Howland, Swift & Allen – were managing agents for an average of ten vessels each.[43]

Whaling agents and their co-owners used a variety of strategies to make their businesses profitable. In contrast to the earliest days of the fishery, when owners kept 25 percent of the catch for themselves (leaving the remainder for the crew), owners now took around 70 percent of the net proceeds. They continued to use the lay system to help offset the risks of the business, paying their whaling crews a specified share (lay) of the net value of the catch after such charges as piloting, wharfing, loading, insurance, and shipping had been subtracted. This meant that if a ship was wrecked or had hard luck on the whaling grounds that the whaling crew, not just the owners, suffered financial loss. And a crewman was entitled to his share of the oil only if he returned to the United States or a prorated share if he was honorably discharged. He was not guaranteed any compensation at all if he deserted.[44]

Merchants continued the practice of rewarding masters with relatively high compensation. In fact, from the early 1840s to the late 1850s, according to economists, they granted masters and mates shorter and shorter lays (larger shares of the profits), while lengthening seamen's lays. Owners increased officers' shares by 12 to 24 percent, while reducing seamen's percentages by 9 to 15 percent.[45]

Shipowners did not rely solely on the sale of the catch to make money. They used the whaleship as a chandlery of sorts. Employing the shipmaster as storekeeper, owners sold clothing and personal items to sailors at sea. Taking profits of up to 100 percent on the items sold (such as tobacco, shirts, and jackknives) owners turned the ship's "slop chest," as this floating store was known, into a very remunerative operation. The shipmaster also took a percentage of the slop chest profits.[46]

Owners also made money by charging seamen interest for loans. They provided

---

[43] "Private Signals of the Whaling Vessels &C Belonging to the Port of New Bedford," Lithograph by Prang & Mayer, 1857, WMLOD.

[44] Byers, *Nation of Nantucket*, p. 93; Vickers, "First Whalemen," p. 567; Butler, "J. and W. R. Wing," pp. 61–2; Hohman, *American Whaleman*, pp. 217–23. Davis et al., "Risk Sharing," pp. 7–11.

[45] Davis et al., "Risk Sharing," pp. 7–11, 58. These scholars suggest that one of the reasons for the lengthening of seaman's lays may be the redistribution of the whaling fleet to the Pacific and Arctic oceans, where men traditionally received longer lays than elsewhere.

[46] Hohman, *American Whaleman*, pp. 246–52, gives a detailed account of the slop chest. He explains that owners considered the 100 percent profit rate "by no means extortionate." Because they assumed the risk of sailors' deserting before seamen had repaid advances, and because these profits were calculated over a *voyage*, not annually, owners regarded them as a reasonable rate of return.

sailors with advances for their sailing outfits (priced at around $75.00) and attached to this an interest rate of up to 25 percent for the voyage. They also authorized captains to lend sailors cash to spend on shore, with an interest rate of 25 percent (again for the voyage). They also loaned money to the families of officers, with a one-time interest rate of up to 40 percent.[47]

These practices paid off. The average level of the industry's profits from 1816 to 1905, according to a recent economic analysis, was 45 percent per year. Even though owners might lose one vessel in ten to disaster and individual voyages might be failures, merchants shouldered these losses by sharing owner-ship widely. And as late as 1900, when the industry has been described as being on the skids, its owners assuredly were not. Many of the businessmen still involved in whaling were earning nearly 70 percent annually on their investment.[48]

III

Historians who measure the peak of things by the numbers of ships and the numbers of men at sea can point decidedly to the end of the 1840s and begin-ning of the 1850s as the height of American commercial whaling. Between 1847 and 1860 (with the exception of two years) no fewer than 600 American whaleships were employed annually on the world's oceans. To say that the United States dominated the world whaling fleet at this time is something of an understatement: in the late 1840s more than 700 of the 900 vessels engaged in whaling worldwide were American.[49]

Within a decade of reaching its zenith, though, the industry contracted by 70 percent, so that in the decade following the Civil War it stood at the same size that it was in the early 1830s. Between 1855 and 1906, with few exceptions, the size of the whaling fleet dropped almost 20 percent every five years.[50]

Students of the industry continue to examine and debate the factors that brought about such a fast decline. Some historians maintain that changes in foreign markets, particularly in England, Germany, Holland, and Belgium, played a role in the shrinking of the industry. When foreign demand for whale oil lessened after midcentury, it delivered an enormous blow to the fishery.[51] More commonly, though, historians have cited the loss of domestic markets:

[47] Lisa Norling, "Contrary Dependencies: Whaling Agents and Whalemen's Families 1830–1870," *Log of Mystic Seaport* 42 (Spring 1990): 4; Hohman, *American Whaleman*, pp. 256–8.   [48] Davis et al., "Productivity," p. 13.

[49] Hohman, *American Whaleman*, pp. 41–2; Moment, "Business of Whaling," p. 263.

[50] Teresa D. Hutchins, "American Whale Fishery," p. 65.

[51] Moment, "Business of Whaling," p. 265; Teresa D. Hutchins, "American Whale Fishery," pp. 272–90. Exports declined in European countries due in part to the rising competition of mineral oils.

first, the diminishing of the illuminant market in the mid-1800s, and then, just after the turn of the century, the loss of the baleen market. With the development of spring steel in 1907, women who continued to corset their bodies discovered that they no longer needed whale's baleen to do it but could use cheap, flexible steel stays instead.[52]

Historians have also attributed whaling's decline to the growing scarcity of whales. Recent studies of whale populations and of cetacean procreation indicate that bowhead populations among others were diminished by overhunting, making the search for and capture of these animals more difficult and more expensive. The effect of whaling on sperm whale populations has been harder to determine, but evidence points to a similar reduction in numbers.[53]

In addition to changing markets and a dwindling stock of whales, a number of other developments helped to extinguish the whaling business. The Civil War not only meant a big boost for whale oil's competitor, petroleum, but prompted events that significantly damaged whaling property. Captain James I. Waddell, master of the Confederate Navy's *Shenandoah*, had the dubious honor of obliterating much of the Arctic whaling fleet at the end of the Civil War. In 1864 he was given orders by Confederate officials to cruise into the "far distant Pacific," to look carefully for (Union) whalers, and to "greatly damage and disperse that fleet."[54]

Waddell did his best to follow orders. Having located part of the Arctic fleet in June 1865 (he said he did not know the war was over), he burned twenty-five whalers to the water's edge. Other Confederate raiders, particularly the *Alabama*, were busy, too, and altogether they eliminated forty-six out of 186 active whaleships. Forty other whalers – all older vessels – were sold to the United States government to be sunk in Savannah and Charleston harbors as an attempted blockade of those ports.[55]

Civil War damage extended well beyond the actual destruction of vessels. In an effort to avoid high insurance rates or to prevent destruction at the hands of the Confederate navy, many whaling owners sold their ships or registered them abroad. This "flight from the flag" meant a further reduction of the fleet. And

[52] Starbuck, *American Whale Fishery*, p. 110; Bockstoce, *Whales, Ice, and Men*, p. 335.

[53] "The Whaleman's Shipping List and Merchant's Transcript," January 6, 1852; Michael James Maran, "The Decline of the American Whaling Industry," (Ph.D. diss., University of Pennsylvania, 1974), p. 33; Teresa D. Hutchins, "American Whale Fishery," p. 343; Richard Kugler, "Historical Records of American Sperm Whaling," in *Mammals of the Seas* III (Rome: FAO of the U.N., 1981), pp. 321–6.

[54] U.S. Secretary of the Navy, *Official Records of the Union and Confederate Navies*, 1st Series III: 749, quoted in Bockstoce, *Whales, Ice, and Men*, p. 106.

[55] Robert G. Albion and Jennie Barnes Pope, *Sea Lanes in Wartime: The American Experience, 1775–1945* (Hamden, CT: 1968), pp. 148–73; Bockstoce, *Whales, Ice, and Men*, pp. 113, 127; Hohman, *American Whaleman*, p. 291.

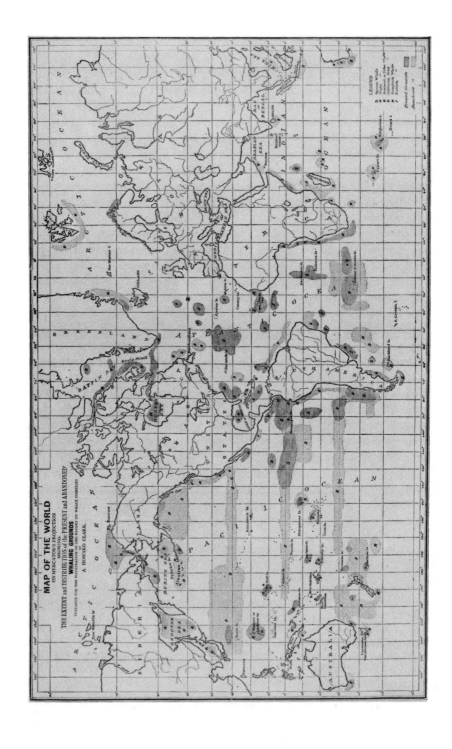

MAP OF THE WORLD
ON MERCATOR'S PROJECTION
SHOWING
THE EXTENT and DISTRIBUTION of the PRESENT and ABANDONED
WHALING GROUNDS
PREPARED FOR THE ILLUSTRATION OF THE HISTORY OF WHALE FISHERY
BY
A. HOWARD CLARK.

because a congressional act in 1797 banned repatriation, vessels sold overseas during the war could not legally be returned to American registry.[56]

Those investors who decided to finance industries other than whaling after the Civil War, as many did, may have felt vindicated by events in 1871 and 1876. In late August 1871, almost the entire Arctic fleet – thirty-three vessels – was lost when pack ice developed earlier than expected and closed in on whaleships. Crews abandoned ship and sailed to Honolulu in whalers that had escaped damage. The annihilation of this fleet, said historian Elmo P. Hohman, "dealt the fishery a body blow." If so, it was not the last hit. In 1876, tragedy struck again in the Arctic, this time with fatalities. Similar weather conditions took out twelve more vessels and this time killed fifty-five men.[57]

Whaling owners may have faced difficulties beyond changing markets, wartime disruptions, and natural disasters. Elmo P. Hohman, among others, claimed that merchants also struggled with the diminishing competence of their whaling crews. During the middle decades of the 1800s, he contended, when "individuals from every race and from a score of nationalities rubbed shoulders in the crowded forecastles and steerages," there was a "marked deterioration in skill, experience, efficiency, and morale."[58] Recent historians of the industry have scrutinized Hohman's assessment and have agreed that widening opportunities lured native-born whites away from deepwater work and that as a consequence whaling owners increasingly hired men who were illiterate and inexperienced. One study maintains that on the eve of the Civil War "almost three-fourths of the deckhands had never been to sea." It also notes the simultaneous decline in crew "quality," which it measures by literacy and skill and the falling off of voyage productivity.[59]

Were shipowners thus dealt another "body blow" by having to hire men who did not know how to whale or men who could not read or write? Probably not. Inexperience may have presented owners with as much of an opportunity as

---

[56] George W. Dalzell, *The Flight From the Flag: The Continuing Effect of the Civil War Upon the American Carrying Trade* (Chapel Hill: University of North Carolina Press, 1940), pp. 248–50; Bockstoce, *Whales, Ice, and Men*, p. 105; James M. Morris, *Our Maritime Heritage: Maritime Developments and Their Impact on American Life* (Washington: University Press of America, 1979), pp. 196–7.

[57] Hohman, *American Whaleman*, 294.   [58] Ibid., p. 48.

[59] Davis et al., "Risk Sharing," pp. 27–9.

(*Opposite*) Those areas of the oceans where whales commonly fed, mated, or calved, as well as the migration paths to and from these areas, were known as "grounds." Naturally, a whaleship sought out whaling grounds in which to conduct its hunt. This illustration, published in a government report of the fishing industries, is one of the few charts of the predominant grounds in existence. The darkly shaded areas are listed in the key as "present" grounds, the lighter areas as "abandoned" grounds.

a liability and may have been something of a bargain.[60] The lengthening of whalemen's lays at the same time that the industry took on more unskilled men is certainly not a matter of coincidence.

In addition, there was nothing new about inexperienced or illiterate men in whaleships, and therefore it would be hard to argue that men like these hastened the industry's decline. Whaling merchants began to recruit verdant farm boys in the late eighteenth century, just as the industry began its meteoric rise. And whaling owners began the fishery in the colonial period and met their first business successes by relying on the labor of presumably illiterate Indians.[61] Explanations for the industry's decline, therefore, are better focused on factors like changing markets or diminishing numbers of whales rather than on the evolving "quality" of whalemen.

[60] Davis et al., "Risk Sharing," pp. 30–2. Davis et al. also suggest that changes in ship technology, which made the ship easier to handle, may have offset greenhand inexperience. For other discussions of the advantages of a green crew, see Charles Nordhoff, *Whaling and Fishing* (New York: Dodd, Mead, 1895), p. 12; Hohman, *American Whaleman*, p. 91.

[61] Daniel Vickers deserves credit for pointing out in conversation, the ironies of ascribing whaling's demise to the very sorts of individuals responsible for its rise.

2

"'Tis Advertised in Boston"
The Shaping of a Ship's Crew

I

The tale may be a familiar one.[1] The hapless mariner, weather-beaten and work-worn, arrives in port after a long sea voyage. He takes sweet, intoxicated relief in the brothels, the beer halls, and the boarding houses of sailortown, but there he gets used. Prey to shady figures of the waterfront, he is soon stripped of his small earnings. His "friends," though – boardinghouse keepers, shipping agents, and outfitters – find a way for him out of debt. With the help of liquor, they ease him into familiar employment. And when he sobers up he finds he is aboard a moving vessel, bound to sea once more.

In certain maritime trades and and in certain ports, such a story has some facts behind it.[2] Evidence from crew lists, shipping agencies, and sailor records tells us, though, that the resemblance it bears to American whalemen and to the process of recruitment in New England in the mid-nineteenth century is only slight. First and most obviously, whalemen (or would-be whalemen) were not pitiably helpless. Prevoyage desertion rates alone speak to their decisive willingness to stand up to their prospective employers. Second, owners of whaling ships operating out of ports like New Bedford were not desperate enough for men to have engaged in wholesale kidnapping or to have sponsored it. Third,

[1] Elmo P. Hohman, *The American Whaleman* (New York: Longmans, Green, 1928), p. 15, offers one version of this scenario.
[2] See, for example, the history of recruitment in San Francisco: Lance S. Davidson, "Shanghaied! The Systematic Kidnapping of Sailors in Early San Francisco," *California History* 64 (1985): 10–17. The recruitment of whalemen in foreign ports or islands or in California may have involved considerably more physical coercion than in northeastern America. See Chapter 6.

the image of a man ensnared in continuous voyaging is not supported by demographic facts. Any man who repeatedly made the rounds of ship and shore, and who sailed on voyages of three to four years would have aged. Whalemen tell us that the aged foremast hand in the nineteenth century was a rarity, and crew lists and secondary evidence indicate that the men who departed from shore in American forecastles throughout the middle years of the nineteenth century were young, usually under twenty-four. Evidence also suggests that many of these sailors were new to whaling (by one estimate, 50 percent of them).[3] The man who was carried into the ship's forecastle against his will and forced into repeated voyaging again and again was, with regard to this industry at this time, a fiction.

If shipowners were not desperate when they hired whalemen and prospective sailors were not duped or stupefied, it remains to be seen what did motivate recruiters and would-be whalemen. The correspondence of shipowners is helpful in this regard, for it reveals a good deal about the sort of men merchants sought for whaling and the sort of men they found during the industry's peak. Letterbooks of New Bedford owners are particularly instructive.[4]

Managing owners, or ships' agents, were generally charged with selecting suitable officers, and according to these letterbooks, they sometimes faced a formidable task. The industry had grown so large by the mid-1800s that an owner could no longer simply call on family or acquaintances for recommendations of reliable men. He might still find a master from a familiar pool of local talent, but he might have to look farther afield for mates. Agent Jonathan Bourne had to cast his reference net wide in the search for officers and wrote to agents and shipmasters in New York, New London, and Boston for information on applicants. He looked for masters who could match specific branches of the industry (a sperm whaleman would not do for the northern fishery, for instance), and he looked for men whom he could trust absolutely. He knew from hard experience, too, what risks he took when he hired men he did not know. Captain Frederick A. Weld had been a stranger to Bourne when the agent gave him a 400-ton ship to command in 1859. Weld had come with a good

---

[3] See Appendix II. Early American merchant seafarers were similarly youthful. See Ira Dye, "Early American Merchant Seafarers," *Proceedings of the American Philosophical Society* 120 (October 1976): 336; and idem., "Physical and Social Profiles of Early American Seafarers 1812–1815" (paper presented at the Jack Tar in History Conference, Halifax, Nova Scotia, October 1990). On the percentages of greenhands, see Hohman, *American Whaleman*, p. 58.

[4] Shipowner correspondence reviewed for this chapter includes the incoming correspondence and letterbooks of Jonathan Bourne, Jr., 1850–61, in "Jonathan Bourne, Jr., Business Records," Mss 18, S–G, S–s 2, vols. 1–3, WMLOD; Charles R. Tucker, 1836–37, NBFPL; Charles W. Morgan, 1833, NBFPL; Swift & Allen, 1849–63, in "Swift & Allen Collection," Mss 5, S–g 3, S–F, S–s 1, Folders 6–76 (incoming) and S–s 2', vols. 1–4 (letterbooks), WMLOD; George Howland, Sr., 1836–47, in "The Howland Collection," Mss 7, S–g 1, S–D, vols. 3–4, WMLOD.

recommendation, but Bourne had not known the referent personally. The captain, it turned out, had a drinking problem, and by the time he reached the Western Islands, he was "insane." Bourne blamed his troubles on the unreliability of Weld's sponsor.[5]

Owners troubled themselves about the trustworthiness of references, but they did not worry much about the numbers of applicants. From their perspective, there was no serious shortage of requests for officer appointments in the mid-1800s – particularly in depression years and during the Civil War, when outgoing vessels were few.[6] The average earnings of whaling captains and mates were greater than those of skilled workers ashore, and this may have helped to swell the ranks of men at this level.[7] As a consequence of a stable labor supply, managing owners could continue to select men of their own sort – white native New Englanders experienced in whaling – for their favored positions.

The managing owner devoted his own time to the selection and recruitment of officers. He handed the direct selection of seamen, though, to a shipping agent. We know too little about these agents – where they were actually based, how they advertised, and how the details of shipping operated. Elmo P. Hohman claimed that many shipping agents were located in the developing regions near the Great Lakes and in cities where young men may have been looking for work.[8]

Evidence from crew lists suggests that it is likely that recruiters did go farther and farther from shore ports to complete a ship's crew. Lists show that men from coastal New England continued to dominate the outgoing forecastles of whalers through the middle decades of the nineteenth century but that they increasingly shared this space with men who hailed from other northeastern regions.[9]

Although the widening geography of recruitment indicates that owners needed to move beyond coastal centers to crew their vessels, it does not mean that they faced a crisis in finding applicants. Owners' letters to captains, to agents, and to other businessmen describe a steady line of men moving through shipping offices and eventually into whaling ports. During the depression years of 1837 and 1857 and the hard times of the early 1840s, especially, New Bedford owners

[5] Jonathan Bourne, Jr. to N. & W. W. Billings, 8 June 1850; Bourne to Seth Coffin, 14 March 1857; Bourne to John A. Baylies[?], 12 September 1857; Bourne to John N. Penson & Co., 14 September 1857; Bourne to Daniel McPhierson, 31 October 1859.

[6] Swift & Allen to Captain Samuel Prentice, 26 October 1857; Swift & Allen to Captain James E. Stanton, 17 July 1858; Jonathan Bourne to Captain F. A. Weld, 18 July 1859; Swift & Allen to Captain Green, 6 August 1861; Swift & Allen to Mr. Taylor, 23 December 1861; Swift & Allen to Reuben Crapo, 9 November 1863.

[7] Lance E. Davis, Robert E. Gallman, and Teresa D. Hutchins, "Risk Sharing, Crew Quality, Labor Shares and Wages in the Nineteenth Century American Whaling Industry," *NBER Working Paper Series* 13 (May 1990): 1, 22, 61–7.

[8] Hohman, *American Whaleman*, p. 90.   [9] See Appendix III.

seem to have turned men away. Jonathan Bourne, for one, was so selective in the late 1850s that he even asked for letters of reference for boatsteerers and foremast hands. And at one point he was so overwhelmed by requests for employment that he suggested to applicants the alternative of the merchant service.[10]

William Allen, who went whaling in New Bedford in 1842, claimed that New Bedford was thronged with young men who had chosen to go whaling and who had to "stay by till by good luck they [could] get a ship." Shipping offices were so crowded at that time, declared Allen, "that the owners and capts were as difficult about pleasing as a lady buying a new dress." "If young men only knew what was in store for them," he noted in hindsight, "4 hours sleep in 24 and now and then the sight of a savage island the owners would have to be less choice [choosy] in their selections of crews."[11]

When it came to selecting new recruits, owners could be choosy indeed. It seems that some owners turned away experienced hands, favoring the supposed innocent compliance of a man new to the sea.[12] On the other hand, it seems that some merchants tried to avoid hiring minors. Shipowner Jonathan Bourne, for example, was careful to enquire of a Mrs. Abby Read of Worcester, Massachusetts, whether her charge, one John Weston, was yet twenty-one, for he wanted her "consent for him to go." "My practice," he explained, "is never to ship a minor unless with the consent of his parents or guardian."[13] We do not know whether Mr. Bourne's fastidiousness was widely characteristic of the industry (the large number of whalemen who were in their teens suggest that it may not have been), but owners who shipped minors were open to lawsuits from angry parents. Minors were legally bound to turn over earnings to their fathers, and one scholar claims that many court cases evolved out of the shipping of boys without parental permission. Even when a boy deserted, argues Gaddis Smith, fathers were "invariably successful" in collecting wages earned until a twenty-first birthday, and owners were occasionally forced to pay double the wages for underage sailors who returned home with the vessel.[14]

Owners tried their best to select men who would be hardy enough to withstand the rigors of the hunt and looked favorably upon men with bulk and muscles. Was a man of "good size" and "stout," agents asked, when they could not

[10] Jonathan Bourne, Jr. to Abby Read, 9 April 1858; Bourne to Gilbert Hathaway, 14 April 1857; Bourne to Captain Manter, 10 April 1858; Bourne to Edward Howell, 12 June 1858.

[11] Diary of William A. Allen, *Samuel Robertson*, 23 June 1842, Log 1040, WMLOD.

[12] See Charles Nordhoff, *Whaling and Fishing* (New York: Dodd, Mead, 1895), p. 12.

[13] Jonathan Bourne, Jr. to Abby Read, 9 April 1858. On the subject of shipping minors see also Lisa Norling, "Contrary Dependencies: Whaling Agents and Whalemen's Families 1830–1870." *Log of Mystic Seaport* 42 (Spring 1990): 4.

[14] Gaddis Smith, "Whaling History and the Courts," *Log of Mystic Seaport* 30 (October 1978): 76. Smith's study includes an examination of cases reported in federal district and circuit courts before 1880.

interview a man in person. Potential recruits, for their part (or those who represented them) stressed their heftiness. George Ketchum, superintendent of the New York House of Refuge, wrote to the agency of Swift & Allen at the end of May 1849 to offer them a number of young men who were "well behaved." But Ketchum also took pains to stress their size and noted their weight as one of their assets.[15]

The search for stoutness was not always successful. Captain A. G. Goodwin of the *Tuscaloosa* reprimanded owner Frederick Allen in 1844 for having hired men deceptive in their size. He claimed that several members of his crew, once they had got to sea, looked no larger than "a dirty shirt" with the "starch all out of them." The hardest case, he lamented, was one individual who had appeared "normal size" when they shipped him but who seemed to have shrunk. "Now I'll let you into the mystery, so that you can guard against it next time," Goodwin cautioned. "He must have had on several suits of clothes as I observed that for several days he kept looking smaller & slimer at last thinks I where will this all end. have we got the walking skeleton on board or not. why he is nothing but *moonshine*."[16]

The complaints of Captain Goodwin and others notwithstanding, the biggest challenge facing owners with regard to recruitment was not finding men to form a crew or shipping them but seeing them weigh anchor. There was a long distance, both temporally and psychologically, between a man's visiting a shipping agent's office, his signing articles (the employment contract), and his actually sailing. A good many men had second thoughts about their prospective voyages, and they successfully acted upon them by running away. Two agents' memorandum books suggest the attrition rates of new recruits. One account shows that around 20 percent of men handled by a New Bedford firm between 1854 and 1872 abandoned plans for whaling and left the city. Another set of placement records from 1847 to 1854 indicates roughly the same percentage of men who did not ship out.[17]

It was the recruiting agent, rather than the owner, who bore the burden of sailor attrition. The recruiter was paid a fee that ranged from $5.00 to $10.00 before the Civil War if he placed sailors aboard a vessel but only a fraction of that figure if he lost a man.[18] In the 1847 to 1854 account, for example, the shipper frequently earned $8.25 a head for a placement. If a man left, however, he kept only the 25¢. Irritated no doubt by his loss rate, this recruiter left wry

[15] George Ketchum to Swift & Allen, 26 May 1849.

[16] A. G. Goodwin to Swift & Allen, 20 December 1844.

[17] Anonymous, "Mss. Memorandum Book of a New Bedford Firm, Giving List of Men Shipped on Whaling Voyages, 1854–1872," described and evaluated in Hohman, *American Whaleman*, p. 93; Anonymous, "Labor Procurement Account Book, 1847–1854," PC.

[18] Hohman, *American Whaleman*, p. 91, says that during the more difficult hiring times of the war and post-war years, recruiters were paid $10 to $20 for each man they placed.

remarks in his book describing the whereabouts of his charges: "absconded," "got scared," "gone to the devil."[19]

Recruiters were irritated not only by uncooperative men but also by the pickiness of shipping agents. One recruiter was paid only if he found men "what we can use." And even when he found proper applicants, he faced the possibility that ships were already manned. Seven percent of men rounded up by this same individual were turned away because all vacancies had been filled.[20]

From the information we have on the selection of whalemen in northeastern America, then, two circumstances stand out: first, ship owners do not seem to have faced a serious shortage of applicants at any time during the mid-1800s, and they occasionally had plenty of men knocking on their doors for work. Second, those applicants might have been numerous and eager to whale, but they were also capable of changing their minds and acting swiftly in their own self-interest. There were, no doubt, some men in the American whale fishery who matched the mythic picture of the hapless seaman, and some shipping agents so desperate to fill a vessel that they shanghaied sailors, but these individuals are not easy to find in the historical record.

II

At the end of February 1834, a man named Richard Boyenton took his pen in hand and calculated the money he had made thus far on his whaling voyage to the Pacific Ocean. According to his careful estimate, Boyenton had earned $6\frac{1}{4}$ cents for five months' work. He was not sure how he might spend this bountiful amount. "I have not as yet concluded," he remarked, "weather to give this as a donation to the sabath school union or to the education foreign mission or Temperance societies. I think However it ought to be the temperance society for we seam as it regards whaleing to have come under the prohibition."[21]

Richard Boyenton and his shipmates aboard the *Bengal* were unusually unfortunate. But when they met up with other whalemen, they must have found some company in their misery. Recent studies of whaling wages show us that the earnings of whalemen before the mast compared poorly to the earnings of unskilled army recruits ashore, farm laborers, unskilled factory workers, and, not least significantly, merchant seamen.[22]

Because men might have earned better money in other trades and enjoyed certain social amenities besides, it is not immediately clear why they chose to go

---

[19] Anonymous, "Labor Procurement Account Book, 1847–1854."

[20] Hohman, *American Whaleman*, p. 93.

[21] Diary of Richard Boyenton, *Bengal*, 28 February 1834, EI.

[22] Davis et al., "Risk Sharing," pp. 1, 22, 61–6; Stanley Lebergott, *Manpower in Economic Growth: The American Record Since 1800* (New York: McGraw-Hill, 1964), pp. 530–40.

whaling. And sailors do say that the decision to join a whaleship was a matter of choice. This does not mean that stark necessity did not play a part in their actions, but they were not comatose when they signed shipping articles. Nor were they kidnapped. Disillusioned whalemen were willing to hold others responsible for the miseries of sailing itself, but they held themselves accountable for having made the first mistake. "I have none to blame bot myself," mourned one sailor in 1866. "I would go a whaling so take it and no grumbling," remarked another. A third man was his virtual echo: "I have put my foot in it & I have got to put up with it."[23]

What attracted these men to whaling work was a vast array of needs and wants. Like sailors of the late 1700s, many nineteenth-century recruits went whaling to push their luck.[24] The work may have paid poorly on the average, but the potential was always there, however slim it might actually be, for a future bonanza. A man might know of hundreds of disillusioned ex-whalemen who had come home to measly sums of money, but what would catch his eye would be the rare tale of the giant catch, the story of the full ship brimming with oil, the reports of men paid off in hundreds of dollars. He had, perhaps, heard of voyages like that of the *Watchman*, of Nantucket, which in 1858 found four barrels worth of ambergris,[25] which alone sold for $10,000; or of the *Montreal*, which, after a voyage of thirty-two months in 1853, sold a cargo worth $136,000; or perhaps of the *Pioneer*, whose cargo was worth $150,000 in 1865.[26] With such stories before them, men could put blinders on against the facts of disappointment.

[23] Diary of George Bowman, *Governor Troupe*, 21 May 1866, PPL; Diary of Charles W. Chase, *Smyrna*, 1 November 1854, Log 655, WMLOD; Diary of Abram G. Briggs, *Eliza Adams*, 25 January 1872, Log 940, WMLOD. These sailors point to the presence of choice in the decision to sail, and social and economic historians explain why they might have made such decisions. Joseph F. Kett describes young white men in rural New England in the 1840s as having "abundant opportunities." Kett, "Growing Up in Rural New England, 1800–1840," in *Growing Up in America*, Harvey J. Graff, ed. (Detroit: Wayne State University Press, 1987), p. 183; Bruce Laurie has pictured growing northeastern cities, expanding with the industrial revolution, as places in midcentury where "jobs were available and efforts at finding work were usually rewarded." Laurie, *Artisans into Workers: Labor in Nineteenth-Century America* (New York: Noonday Press, 1989), p. 27. Davis et al. have argued that native-born men in the 1840s and 1850s might indeed have found shore-based opportunities that paid them more than whaling work. Davis et al., "Risk Sharing," p. 22.

[24] Daniel F. Vickers, "Nantucket Whalemen in the Deep-Sea Fishery: The Changing Anatomy of an Early American Labor Force," *Journal of American History* 72 (September 1985): 295.

[25] Ambergris is a wax-like intestinal excretion of whales that is usually found in very small quantities. It was in high demand in the perfume industries as a bonding agent, and according to Elmo Hohman, it was valued in Asia as an aphrodisiac. Hohman, *American Whaleman*, p. 4.

[26] Alexander Starbuck, *History of the American Whale Fishery* (1877; reprint, Secaucus, NJ: Castle Books, 1989), p. 148.

To listen to the ambitions of new whaling hands is to hear their hopes for quick fortunes. Men aboard the *Columbus* in 1852 left home, said one journal keeper, "with the Expectations of makeing a fortune in a short time." Other seamen spoke of the "wealth to be acquired" or their "highest expectations" of a prosperous voyage. When they dreamed of alternatives to whaling schemes in the 1850s and 1860s, they did not picture farm labor or city employment; they fantasized instead about the "gold diggings." "One year of our voyage passed and no oil," noted a discouraged Edwin Pulver in 1852, "I think of trying the gold digins the first chance I have to get to where they are." "A great many" of the crew of the *Julian* in 1848 had similarly "caught the gold fever." And virtually the whole crew of the *William Wirt*, according to one seaman, "are highly excited concerning California [and] are determined to go there."[27] These men, like so many others who headed off to the hills of California or to Australia at midcentury, wanted a fast, certain resolution to economic woes.

These whalemen dreamed of future loot, but they did not at the same time crave a life of material splendor. What many of them most wanted from their voyages – and here again they echo their predecessors of a century earlier – was economic independence and the means for the "necessities of life" – a competence in other words. If they were lucky, the whaling lay would offer them the chance to buy a farm or the wherewithal to attract a wife.[28]

Whaleman Jared Gardner noted in 1841 that he was "quite shure that I shall get enuf . . . to turn my attention to farming." Similarly, Ambrose Waldron was a young man thinking "About bying A farm." Lewis Williams and his shipmates on the *Gratitude* fended off monotony by spinning tales of future financial security: "All hands," he reported, "are going to either Turn farmers or Storekeepers the latter seems to be the Favorite if they ever do land in [sweet] Amerika."[29] The phrase most frequently repeated in these journals was "Who would not sell a farm and go to sea?" and it was patently ironic in its intent.

Having the resources to purchase land also meant being able to provide for a wife and family, and for many young men this was the primary social goal of the voyage. One whaleman was "a poor man that is thinking of getting married when he gets back if he is so fortunate." Another hoped to get "spliced off the strength of this voyage" and a third imagined "payday" when "there will be more marriages take place and more farms bought than was Ever heard of before."[30]

[27] Diary of Edwin Pulver, *Columbus*, 31 January 1852; 4 September 1852, PPL; Diary of F. Cady, *Julian*, 20 September 1848, KWM; Diary of Samuel Allen, *William Wirt*, 17 November 1849, PMS.   [28] Vickers, "Nantucket Whalemen," pp. 294–6.
[29] Jared Gardner to Harriet Gardner, June 1841, AAS; Diary of Ambrose Waldron, *Bowditch* 1 June 1847, PPL; Diary of Lewis Williams, *Gratitude*, 8 April 1859, PMS.
[30] Chase diary, 27 June 1854; Diary of Ezra Goodnough, *Ann Parry*, 24 July 1847, PMS; Diary of John Joplin, *Ann Parry*, 25 April 1848, PMS.

Would-be sailors were hardly the only northeastern men hoping for a competence, looking for work, and consequently on the loose in the mid-nineteenth
century. The circumstances that sent these men from their farms, villages, and
towns to shipping offices were in part the same phenomena that sent men into
unsettled areas of the Midwest, that pushed them across Panama and up the
California coast, and that drove them into cities by the thousands. Opportunities for property ownership and the economic security that went with it were
diminishing at home. Land, in particular, was harder and harder to inherit and
less available to buy cheaply.

In southern New England a "land crunch" was by no means a new phenomenon, and it had played a long-standing role in moving men away from
home and into new lines of work. Beginning in the early to mid-1700s there
were decreasing possibilities for freeholding. Population growth, together with
repeated division of land, meant that many eighteenth-century sons could not
support a family on their fathers' property and had to look elsewhere for land
or an income. Many of them took jobs for wages in commercial towns and
worked in mills and shops or served as farm laborers in the hope of earning
sufficient funds to pay for farms or businesses of their own.[31] Alternatively, would-
be farmers went into the unpopulated areas of Maine, western Massachusetts,
and New Hampshire. There in the rocky outback, they looked for property they
could call their own. And some of these New England colonists, to improve
their chances for independence, went whaling.

The land squeeze that began loosening men from the village economy in the
eighteenth century was still sending men from their New England homes 100
years later. But this time, they could go greater distances. A transportation
revolution had created a network of canals, railroads, and highways that linked
hinterland farms to markets, and hopeful farmers pressed past the Appalachian
Mountains into Indiana, Illinois, and, especially after the news from Sutter's
Mill in 1848, well beyond. At the same time they also stepped up their migration
into cities. Between 1820 and 1860, seaports, riverside towns, and lake-based
towns grew by 800 percent.[32]

When young men disengaged from their families, which at the beginning
of the nineteenth century they did at an earlier and earlier age, they faced a

[31] James A. Henretta notes that "as early as 1771, 29% of the taxable residents of Boston were
neither property owners nor dependents of a propertied member of the community." See
The Evolution of American Society, 1700–1815: An Interdisciplinary Analysis (Lexington,
MA: D. C. Heath, 1973), p. 195. For descriptions of the land squeeze in colonial Massachu-
setts, see Philip J. Greven, Jr., *Four Generations: Population, Land, and Family in Colonial
Andover, Massachusetts* (Ithaca: Cornell University Press, 1970); Robert A. Gross, *The
Minutemen and Their World* (New York: Hill & Wang, 1976).

[32] Bruce Laurie, *Artisans into Workers*, p. 25. For the Boston case, see Peter R. Knights, *The
Plain People of Boston: A Study of City Growth* (New York: Oxford University Press, 1971).

daunting and unfamiliar world.[33] But they also looked at a future freighted with opportunity. In fact, men were not invariably pushed from the thresholds of family farms to seek their way reluctantly in the cold unknown. Some men left home enthusiastically. Other men ran away. Viewing family life, parental control, or perhaps an apprenticeship more as confinement than as security, they packed quickly, left quietly, and made an unequivocal break for freedom.[34] In the records of the whaling industry we see evidence of their flight. Shipowners devoted considerable time writing to guardians, employers, and parents who were looking for young charges. Owners offered them small solace. They told parents that they might contact their truants at a whaleship's first landfall – usually Talcahuano, Chile, or possibly in the Hawaiian Islands. Perhaps the parents would be lucky, they added, and the sailor in question would return in two to three years. Owners also made it clear that concerned parents or guardians had plenty of company. "I should think it verry probable that he sailed from this place," wrote merchant Charles Tucker about a boy named Joseph Warren. "It has of late become verry fashionable for sailors to assume some fictitious name by which they ship and are known by before they sail and by so doing many deprive their friends from tracing them."[35]

Among seafaring families, a son's decision to sail might not have been so precipitous or dramatic. Nor did it necessarily signal a removal from home and family. Boys who grew up in maritime households were expected to carry on traditional means of providership, which often meant joining a father at sea, perhaps even on the same ship. "Sometimes I wish," young Benjamin Cromwell wrote to his father, Peter, at sea in 1854, "that I was there to help you out a little for I think that I could lern to kill a whale after a while . . . you know that I can catch a bluefish, and a whale is but a little smarter."[36] When the Benjamin Cromwells of the world grew up and learned the difference between the size of

[33] Between 1750 and 1820 the average age of departure from home dropped from the midtwenties to the late teens. See Karen Halttunen, *Confidence Men and Painted Women: A Study of Middle-Class Culture in America, 1830–1870* (New Haven: Yale University Press, 1982), p. 12.

[34] As Daniel Scott Smith has reminded us, childrens' determination to be independent had as much to do with the democratization of politics and with changing religious beliefs as they did with an agricultural crisis and a commercial revolution. See "Parental Power and Marriage Patterns: An Analysis of Historical Trends in Hingham, Massachusetts," in *Growing Up in America*, Graff, ed., pp. 156–69.

[35] Jonathan Bourne, Jr., to Edward C. Fisher, 18 January 1852; John Girdwood, "An Address Delivered Before the New Bedford Port Society on the Occasion of the Twenty-Eighth Anniversary," 28 February 1858. NBFPL. Girdwood, who kept crew records for the Society, remarked that he had received "numerous letters of inquiry after sons and friends gone to sea"; Charles R. Tucker to James Osborne 24 July 1837, NBFPL.

[36] Benjamin Cromwell to Peter Cromwell, 28 June 1854, DCHS.

bluefish teeth and those of sperm whales, they would join many others from their home towns.[37] These men may have looked at whaling not only as a way of following family members to sea but also as a means of rejoining them on land. Once they earned sufficient money, many whalemen planned to buy property close to home and settle with familiar folk. Unlike sons who moved west or north permanently to carve out a distant homestead, these whalemen sailed away, in a sense, to stay nearby.

Whalemen, particularly veterans, identified another dimension to the decision to sail. Their diaries are rife with comments that equate sailing with disengagement from their mothers. Sailor Dan Whitfield noted that among the sailors on his ship were those who left "Mamma's Apron Strings in A Pet." Writer Charles Nordhoff similarly referred to greenhands as men who had "cast loose from [their] mammy's apron-strings."[38] New sailors aboard the *Java* encountered rough seas and heavy winds as they entered the Gulf Stream in October 1857. As they began to be sick to their stomachs, said veteran seaman Albert Buel, they were "a wishing themselfes to home with their mamys." But "they soon got over that," he reported with an air of accomplishment. George Blanchard also boasted that a true sailor was a man who was independent of his mother. In the middle of a vicious storm at sea in 1845, Blanchard comforted himself with the thought that, compared with men at home who were clinging to maternal security, he and his shipmates had had the guts to brave the world on their own. "The poor devils on shore," he remarked, "who cannot muster courage enough to leave their Mammas for a week have all my pity."[39]

Although veterans like these pictured the separation of mother and son as something of a triumph, it is by no means clear that novice sailors saw it in such unambiguous terms. The depth of sorrow expressed and even tears shed soon after departure suggests that some young men went to sea with maternal ties still intact. And this is hardly surprising. Changes in the Northeastern economy along with shifts in religion and politics meant that mothers in the mid-1800s were an expanding presence in families as child-rearers and domestic authorities. More and more fathers left home to work in factories, shops, or offices, and became physically and emotionally distanced from household matters. Although the degree to which a mother took on the solo job as caregiver varied – there were still many fathers at home in farming families, for instance – evidence

---

[37] Appendix III shows that the percentage of men from coastal New England in New London crews ranged from about 40 to 50 percent at midcentury. Examination of a small sample of crew lists contained in diary sources (N = 15) from 1835 to 1871 shows that, on the average, 39 percent of crews (excluding master) hailed from coastal Massachusetts.

[38] Diary of Dan Whitfield, *Dr. Franklin*, c.1856, PC; Nordhoff, *Whaling and Fishing*, p. 28.

[39] Diary of Albert Buel, *Java*, 22 October 1857, PPL; Diary of George Blanchard, *Solomon Saltus*, 18 November 1845, PC–TB.

suggests that in many different households women played larger roles in raising children than did men.[40]

The growing presence of mothers in the lives of young men led to an increase in the "emotional power of the mother–son bond."[41] Yet even as (and in part because) mothers and sons saw a great deal of each other, there was pressure on boys as they grew to assume distinct gender identities. Masculine ideals varied, but one thing was clear for many of the male sex at this time: they were not to appear sympathetic, nurturing, tender, and sentimental like their mothers.[42]

But how might a boy learn to be "unwomanly"? Farm boys or artisans' sons could learn appropriate behaviors from their nearby fathers. But even *they* might look outside the home arena for the frequently tough induction into manhood. Some boys practiced being a man through a distinctive rough-and-tumble "boy culture" or learned about the right sort of behavior in fraternal societies or sex-segregated gangs or social clubs.[43] Certainly many young men learned about male camaraderie and unfeminine behavior the grim way, on the battlefields of Mexico and the Confederate South. And obviously some boys chose to become men at sea.

A boy growing up in antebellum America, even far from the coast, had many ways of learning that the sea was a prime testing ground for male hardihood. American popular culture told him that the ocean produced men of a distinctly courageous stamp. A children's book, *Glimpses of the Wonderful*, published in New York in 1845, is a good example of publications that conjoined seafaring with male ideals of courage and daring. "The South Sea whale-fishery," its author explained, "is an undertaking of great difficulty and danger; and we

[40] Robert L. Griswold, *Fatherhood in America: A History* (New York: Basic Books, 1993), pp. 13–16; Julie Matthei, *An Economic History of Women in America: Women's Work, the Sexual Division of Labor, and the Development of Capitalism* (New York: Schocken, 1982), p. 111; Nancy F. Cott, *The Bonds of Womanhood: 'Woman's Sphere' in New England, 1780–1835* (New Haven: Yale University Press, 1977); Ruth H. Bloch, "Untangling the Roots of Modern Sex Roles: A Survey of Four Centuries of Change," *Signs* 4 (1978): 245–6.

[41] E. Anthony Rotundo, "Body and Soul: Changing Ideals of American Middle-Class Manhood, 1770–1920," *Journal of Social History* 16 (Summer 1983): 30.

[42] See Bloch, "Untangling the Roots," p. 245; E. Anthony Rotundo, "Boy Culture: Middle-Class Boyhood in Nineteenth-Century America," in Mark C. Carnes and Clyde Griffen, eds., *Meanings for Manhood: Constructions of Masculinity in Victorian America* (Chicago: University of Chicago Press, 1990), pp. 15–36; Mark Carnes, *Secret Ritual and Manhood in Victorian America* (New Haven: Yale University Press, 1989), pp. 110–25; Elliot J. Gorn, *The Manly Art: Bare-Knuckle Prize Fighting in America* (Ithaca: Cornell University Press, 1986), pp. 140–2. Exceptions to the unsentimental and unemotional ideal for men did exist. Donald Yacovone identifies an importantly variant set of masculine ideals among abolitionists in the antebellum period. See "Abolitionists and the 'Language of Fraternal Love'" in Carnes and Griffen, eds., *Meanings for Manhood*, pp. 85–95.

[43] See Rotundo, "Boy Culture," and for discussions of fraternal organizations and their function in inducting young men into adult manhood, see Carnes, *Secret Ritual*.

A WHALING VOYAGE

Around the World!

ALL HANDS AHOY! CAPT. GEORGE W. ROWELL,
a Practical Navigator, Sailor and Orator, of Manchester, N. H., will give the most thrilling go-a-head and WIDE-AWAKE lecture ever given from an American stage, at this place,

on his first voyage at sea on a whaling voyage around the world, in 8 different sections or parts:

1st. From his native home to New York shipping office, to the place of weighing anchor and making sail, from thence island to island, headland to headland, and whaling ground to whaling ground, the whole circumference of the globe.
2nd. His first lessons at sea.

3d. The number of kinds whales at sea.
4th. The manner of capturing whales.
5th. The manner of heaving in their blubber.
6th. The trying out oil.
7th. A description of a terrific storm at sea.
8th and last. His most pleasant, blissful and mirthful voyage to SEE.

All persons wishing to hear a true life at sea will surely come. Satisfaction guaranteed or money refunded at the door. **ADMISSION—Adults   Children under 12, 10 cts.**

Broadsides like these, posted in villages and on city streets, advertised the excitement and worldly dimensions of whaling. They helped to lure scores of landlocked young men to sea, where months of drudgery convinced many that they had been hoodwinked.

cannot help admiring the resolution of the brave men who engage in it. They embark in a ship, and sail away to a distant ocean, ten thousand miles away from home, to attack one of the very largest animals in the world."[44]

An impressionable boy might have read such publications. He might also have seen any number of whaling "representations," or lectures, that toured the cities. These lectures described men surviving fights with "monsters of the deep," sailors battling sea storms, and virile males holding forth with women of the world. In one theatrical event, presented in Boston in 1860 to "unqualified praise," whalemen dueled with "piratical schooners" (and won); they faced vicious alligators and tigers in Peru; they comported themselves bravely when a mad whale threatened them by "craunching up a whaleboat," and they engaged

[44] Anonymous, *Glimpses of the Wonderful* (New York: Wiley & Putnam, 1845), p. 58. See also Mark Forrester, ed., *Forrester's Pictorial Miscellany* (Boston: F & G. C. Rand, 1854), and Anonymous, *The Whale and the Perils of the Whale-Fishery* (New Haven: S. Babcock, c.1845).

"savages" in war canoes in the "Feegee Islands." "Liberal arrangements" were made for groups of school children to attend afternoon exhibitions.[45]

Youths may have learned of seafaring through such enthusiastic depictions of adventure and challenge, but they no doubt also learned about deepwater life through more cautionary tales written not by ex-sailors but by ministers and women.[46] These were stories designed to discourage young boys afflicted with wanderlust and to convince them of the advantages of staying put.

What, we wonder, would have been the reaction of a young boy to the short tale of the Russel family, which circulated in 1845? The author, "a clergyman," established his story as a power struggle between the sexes. As a storm brews outside the Russel home somewhere on the New England coast, mother Russel and daughter Anne Russel fret over Father's safety on a returning whaleship. Young William Russel, though, has the temerity to interrupt their fearful conversation with the announcement that he intends to go to sea just like his father. "You are always frightened if the wind blows any," says William to his sister. "When I go to sea, I hope we shall not have a crying girl on board." "But you are not going to sea," Anne retorts.

Anne is, of course, correct in this matter, for William is persuaded, after listening to tale upon tale of marine hardship and peril, to relinquish his ambition. And when he actually renounces his seafaring dreams, he flies "to comfort his mother." Throwing his arms around her neck, he exclaims, "Mother – dear mother – I did not know that whaling was so dangerous."[47] Were such stories a disincentive to sail? Chances are they were also a call, albeit unintended, for defiance and for departure.

For a minority of whalemen, sailing away on a deep-sea vessel meant not only leaving mothers but also abandoning "proper" and "civilized" society and genteel company. Having left homes where men did not do manual work or mingle easily with strangers, these Victorian sailors commonly reacted to their first days at sea by asking "What have I done?"[48]

What, indeed, had these "outsiders" done? Victorian seamen offer several explanations for joining the crew of a deepwater whaler. They were not averse to gambling for riches and to using the voyage to strike out on their own,

---

[45] Broadside, "Seventh Week of Capt E. C. Williams' Celebrated Representation of a South Sea Whaling Voyage!" (Boston, 1860), WMLOD (on exhibit).

[46] Ann Douglas, *The Feminization of American Culture* (New York: Avon Books, 1978), p. 54; Myra Glenn, "The Naval Reform Campaign Against Flogging: A Case Study in Changing Attitudes Toward Corporal Punishment, 1830–1850," *American Quarterly* 35 (Fall 1983): 410.

[47] "Stories About Whale-Catching and the Toils and the Perils of the Whalers," in Thomas Teller, ed., *Teller's Amusing and Instructive Tales* (New Haven: S. Babcock, 1845), pp. 4–63.

[48] See Diary of Orson F. Shattuck (under pseudonym of Charles Perkins), *Frances*, 1850–52, Log 994, WMLOD; Diary of Elias W. Trotter, *Illinois*, 1845–47, Log 1005 A & B, WMLOD.

independent of their parents. What they most sought in seafaring, however, was that personal attribute that would qualify them for bourgeois society: self-control. Sometimes this had meaning in medical terms. A number of these sailors sought to better their health in what they saw as the salubrious and ascetic environment of the sailing ship. Most, though, hoped to upgrade their moral condition with a life of strict regimentation. Whereas their middle-class compatriots ashore turned to moral reform organizations like temperance societies for help or to social reformers like Sylvester Graham, these men sought self-restraint in a whaler.[49]

That the whaleship should have been the site of moral regeneration was ironic because mariners in general were viewed by the urban middle class as exemplars of vice and deviance.[50] Yet numbers of upstanding men defied social convention and sought sea captains who were pious and staid, and ships committed to temperance. They also planned to bypass, if they could, the contaminating influence of sordid shipmates. Indeed, the company of strangers could serve to facilitate the self-scrutiny that they sought. Charles Stedman of Norwich, Connecticut, was "sad" thinking of his estrangement from "friends and associations" in America. Yet he was pleased not to be "exposed to all the temptations of former connections." Orson Shattuck, a twenty-three-year-old from New Hampshire, had "grown old in Sin." As a first step toward the reclamation of his morals, he went to sea to "dole out a life of repentance and misery among strangers." Elias Trotter summed up the ambitions of Victorian sailors with his remark that a whaleship, with its "dangers . . . toil & hardship" could teach a great thing: ". . . self denial."[51]

Whaleships were especially sought after as places to remedy alcoholism. Dan Whitfield, a self-labeled "drunkard" who had visited his share of "rum hells" ashore, sent himself whaling in 1856 to dry out. He prayed that he would be able to perform the voyage and could cure himself of his "evil habit." No less prominent a maritime personage than Jane Slidell Perry, the wife of Commodore Matthew Perry, saw a whaler as the way to cure her son William of his drinking. Mrs. Perry wrote to her son's shipmaster that she had sent her son whaling because "he had formed a taste for drinking." She hoped that a stint on a temperance whaler would enable William, who had a "very kind affectionate disposition," to be able to remain at home in the future.[52]

[49] Stephen Nissenbaum, *Sex, Diet, and Debility in Jacksonian America* (Westport, CT: Greenwood Press, 1980), pp. 142–3.

[50] For discussions of sailors' social reputations ashore, see Glenn, "Naval Reform Campaign," p. 410; Hugh Davis, "The American Seaman's Friend Society and the American Sailor, 1828–1838," *American Neptune* 34 (January 1979): 45.

[51] Diary of Charles Stedman, *Mt. Wollaston*, 4 July 1854, NBFPL; Shattuck diary, 27 October 1850; Trotter diary, 19 October 1845.

[52] Whitfield diary, November 1856; Jane Slidell Perry to Capt. S. W. Fisk, 27 October, c.1846, MSM.

Many Victorian seamen looked to seafaring as the way to achieve self-restraint in their behavior. Some also hoped to find exhilarating physical challenges on deep water. Unlike the majority of men surveyed in this book, who were used to hard work at home, on the farm, in the factory, or in a shop, these men were strangers to muscle strain. Often fresh from academies or college or clerical jobs, they hungered for what an American president would later call "the strenuous life."[53] William Abbe, recently of Harvard, anticipated with pleasure a "long & adventurous voyage . . . & full of exciting sport." Robert Weir, whom officers referred to as a "gentleman in disguise," gallantly announced to himself that the "beginning of independence is labor," and he reveled in the sight of his hands as they became hardened by physical work: "I am getting quite used to work now, and my hands can testify to that quite plainly – for they are as hard as horn inside – pulling and hauling on hard ropes – and the outsides have a most beautiful brandylike brown color . . . the handling of ropes & tar has a very visible effect on the hands. I like the sea."[54]

The ocean itself drew many of these men to the recruiting office. The early nineteenth century was an age of Romantic sublimity, when genteel men and women thrilled at the prospect of seeing God in mountainous waves, awe-inspiring storms, and brilliant skies. It was a time when many middle-class Americans began to take vacations to "sublime" locales. Tourists with sufficient leisure time and ready money packed up their sensibilities and headed to sites such as the White Mountains, the Catskills, and above all Niagara Falls. There they experienced the simultaneously awful and beautiful magnificence of nature.[55] Men with a greater sense of daring could look for sublimity in that place where Byron and Coleridge and other proponents of the aesthetic said it resided most decidedly: at sea.

There were whaling recruits, then, for whom the ocean signified Romantic potential and self-improvement and others for whom it held social and economic promise. Although these men differed in their basic hopes for a voyage, they did have a common goal. They all went sailing to see a wider world. Tired of northern winters and bored by hometown landscapes, they listened raptly to the depicted idylls of life in warm settings, and they let their fantasies dance

[53] On Teddy Roosevelt and the "strenuous life" see Kathleen Dalton, "Why America Loved Teddy Roosevelt," in *Our Selves/Our Past: Psychological Approaches to American History*, Robert J. Brugger, ed. (Baltimore and London: Johns Hopkins University Press, 1981), pp. 276–7.

[54] Diary of William A. Abbe, *Atkins Adams*, 11 October 1858, Log 485, WMLOD; Diary of Robert Weir, *Clara Bell*, 29 August 1855; 31 August 1855; 23 September 1855, MSM.

[55] John F. Sears, *Sacred Places: American Tourist Attractions in the Nineteenth Century* (New York: Oxford University Press, 1989), pp. 4–10; Elizabeth McKinsey, *Niagara Falls: Icon of the American Sublime* (Cambridge, UK: Cambridge University Press, 1985).

with the thought of faraway lands. Advertisements that promised "beautiful countries to be seen" and "mild climates" spoke with magnetic allure, and drew seaward hundreds of landlocked men.[56]

James Cather was susceptible to such attractions. During a hot midsummer in 1854, this young man found himself on the streets of New York City looking for work. "Not being acquainted with any persons," though, he "could not get a situation." He had been in the city a little over a week when, while passing along South Street, he saw a notice advertising for whalemen.[57] We do not know exactly what the broadside said, but we know what it meant to Cather. After he had applied for a whaling job, been accepted, and sent aboard, Cather notified all his friends and relatives of his itinerary. He was bound, he announced proudly, on a "journey around the world," and in every letter, as if to relish the exotic sound of each separate distant destination, he recited a litany of ports: "St. Helena, Callao, Sandwich Islands, Japan Seas, Indian Ocean, Pacific Ocean, San Francisco, Valpariso & Rio Home."[58]

Seaman George Blanchard would have applauded Cather's efforts to become cosmopolitan. Sailing around the Horn, he argued, was a man's route to respect and envy. "I say shipmate," he asked rhetorically in 1845, "hast [thou] ever doubled the Horn? If not, bring not thy petty experience of the pleasures of the indigo-blue of the Atlantic into contrast with the delights of the light-blue Pacific – No matter how often thou hast been nearly 'chawed up,' in the chops of the Channel; no matter what may have been thy voyages, up the Straits, or on the Coast of Africa; unless thou hast doubled the Horn! lo! and behold, I write thee down an Ass – Heave, and weigh – Top your boom and be off."[59]

For mixed and varied motives – a desire to be more worldly, a need to make some money, a wish to be independent, a yearning for the "sublime" – men signed up to become sailors. In their ambitions, they were not unlike their compatriots ashore who also sought livelihoods, social freedom, and self-betterment. But greenhands stood at the threshold of a professional experience widely regarded as transformative and other-worldly. And in the chapters ahead we will consider the degree to which these men held tight to their shore-given selves and social habits and the degree to which they distanced themselves from "landsmen" and together forged a community apart.

[56] Allen diary, *Samuel Robertson*, n.d. (c.1842).
[57] Diary of James H. Cather, *Roman*, 29 July 1854, KWM.   [58] Ibid.
[59] Blanchard diary, *Solomon Saltus*, 28 December 1845.

3

Wondrous Tales of the Mighty Deep

Whaling Life and Labor

A wondrous tale, could the rare old whale,
Of the mighty deep disclose;
Of the skeleton forms, of by-gone storms,
And of treasures, that no one knows.

He has seen the crew, when the tempest blew.
Drop down from the slippery Deck,
As he shook the tide, from his glassy side,
And sporting 'mongst Ocean and wreck.

Sea song, in journal of George Blanchard
Solomon Saltus, 16 August 1847, PC-TB

I

Perched aloft on a topsail yard in a gale, William Abbe tried to reef topsails for the first time in his life. He found the experience nothing short of exhilarating. "The wind blowing . . . the officers shouting from the deck & the 3rd mate singing out from the Earing – Haul to leeward or to windward – all this and the raised & elevated feeling I experienced from the novelty & danger of my position – made my first trial at reefing topsails the pleasantest part of my day's duties." Abbe, who had brought copies of Coleridge to sea with him, thrilled to the sights and sounds of a wild night at sea. Richard Boyenton, another sailor, would not have sympathized with Abbe's delight. Hanging onto yards in gale winds, facing the awesome dangers of nature, and being ordered about by officers who stayed dry and safe on deck held little charm for this mariner. "The sublime scenery described by some is entirley lost uppon me," he remarked.

"Let [these same writers] come awhaling, and have to earn a living, and these fine flights of . . . fancy would end in the blues."[1]

As the cases of these two men demonstrate, men before the mast responded to whaling work and to life at sea in singular ways. They viewed their experiences through the lenses of dissimilar conventions and according to idiosyncratic needs. Whether whaling was successful, satisfying, onerous, or troublesome depended in part on a sailor's personal ambition. Whaling signified much more, however, than the mirror of a man's past and the reflection of his unique goals. The actual effort of sailing the ship, the catching of whales, and the lengthy sojourn in distant seas fostered among sailors a common set of concerns and a shared sense of distinctiveness.

To understand what whaling meant to men of the mid-nineteenth century and how it may have affected them, it is important to know, step-by-step, what deepwater work involved. This chapter studies the whaling voyage from its very beginning, just after the whaleship weighed anchor in home port. And it starts with that miserable, memorable experience that signified a man's first real introduction to life afloat.

Sometimes that experience took hold of new sailors only hours after weighing anchor. Other times it waited until greasy ocean swells began to roll the vessel from side to side or until a Gulf Stream gale began to pitch the ship. But almost inevitably, seasickness overcame greenhands, slowly at first, and then with violent insistence. Aboard a whaler called the *Atkins Adams*, men began to be sick by the second day, and by the fourth the forecastle was an uninhabitable hole. "The sick men," a sailor wrote, "vomited out on the floor and the vomit ran down behind the chests or collected in heaps on the floor. To this was added bits of meat and bread – onion skins – spilt coffee – tobacco spittle – forming in all a disgusting compound." The sight, the smell, and the sound of straining men made it impossible for this whaleman to rest, and he walked the decks during his watch below – better, he felt, to lose some sleep than the contents of his stomach.[2]

Another greenhand reported that he "vomited till [his] nose bled." A third would have offered his "kingdom for a quiet stomach." Sea captains were occasionally sympathetic to the plight of novices – the shipmaster aboard one vessel made gruel for the men and "cheered them up" – but captains also demanded that the ship sail on and ignored their crews' incapacity.[3] Hardly daunted by the sight of retching men, in fact, masters usually organized a company meeting during the first day at sea, in which they chose watches,

---

[1] Diary of William A. Abbe, *Atkins Adams*, 11 October 1858, Log 485, WMLOD; Diary of Richard Boyenton, *Bengal*, 31 May 1834, EI.   [2] Abbe diary, 10–12 October 1858.

[3] Diary of James Bond, *John A. Parker*, 26 October 1852, NBFPL; Diary of George Blanchard, *Solomon Saltus*, 6 August 1845, PC–TB; Abbe diary, 22 October 1858.

outlined the itinerary for the voyage, and exhorted sailors to work hard and be obedient. On at least one occasion, though, prostrate seamen got the better of a captain's indifference. Captain Goodwin, of the *Tuscaloosa*, felt little pity for nauseated greenhands and was even amused at their expense. Every time he issued an order, he explained, "they would raise their heads & turn up the white of the eye, with such a comic expression, as much to say; 'dont Capt dear dont you see the situation we are in.' " Goodwin did not see the situation apparently, for in the middle of his sailors' ordeal he decided to call them aft "to speechify a little to them." Gathering his crew around him, he launched into his address. He could barely complete a sentence, however, before the loud sound of retching stopped him. "But ever & anon," he reported, "I was interrupted in my speach by a strange *gutteral* sound & an oh dear, then a spasmodic *start*." The noise proved such a distraction that he could not go on.[4]

Having recovered their equilibrium and their appetites and having survived the first real test of their stamina, whaling recruits turned next to the unfamiliar work of the sailing ship. The watch schedule provided one of the most baffling challenges to those inexperienced in sea labor. Watches referred to both groups of men on duty and to the stints of duty themselves. For much of the time that the whaler was making a passage (sailing to and from cruising grounds), the crew was organized into two watches under the command of the first and second mates. These groups worked around the clock, in four-hour stints of duty, except during the dogwatch, a short period of two-hour watches from four to eight o'clock at night. The watch rotation frequently shifted when a ship reached cruising grounds. Then, the crew was further subdivided into boat's crew watches. Headed by boatsteerers instead of mates, these groups divided the work rotations into as many cycles as there were boats' crews and thus gave all sailors more sleep.[5]

If the whaleship sailed from the United States to the Azores or the Cape Verde Islands to add men and supplies, officers used this leg of the voyage to train novices in sailing skills. Foremast hands attempted to quickly learn the lines, the names of sails, and the points of the compass. Time and again they practiced furling and unfurling sails. They took their two hour "tricks" at the wheel, staring out at the long, empty horizon, trying to keep themselves steady and their ships on course.

Sailors also had to learn an array of skills having to do with the whale hunt. It behooved them to know — for their safety depended on it — the detailed mechanics of whaleboats. There were, depending on the size or rig of the

---

[4] Captain A. G. Goodwin to Swift & Allen, 20 December 1844, "Swift & Allen Collection," Mss 5, S–g 3, S–F, Ss 1, Fol. 6–76, WMLOD.

[5] Diary of John Martin, *Lucy Ann*, 16 May 1842, KWM; Diary of James Haviland, *Baltic*, 7 January 1857, PPL; also see Elmo P. Hohman, *The American Whaleman* (New York: Longmans, Green, 1928), p. 153.

vessel, four or five whaleboats slung from davits on the whaleship's sides (three on the port side; one or two on the starboard side), and greenhands would learn that it was in these 28- to 30-foot wooden boats that they would pursue whales.[6] The boats were light, strong, seaworthy craft but not invulnerable. Depending on the weather or the wariness or distance of whales, the six whalemen who manned these boats either paddled, rowed, or sailed. They took with them equipment that would secure and kill whales, such as harpoons (whalemen called them "irons"), lances, spades, and a drag to slow the whale. They stowed at least 300 fathoms of line, neatly coiled in two line tubs.[7]

Whaleboats carried a variety of complicated gear, set up so that there was no mistaking where things were. So much of the equipment was lethal, in fact, that the boats' organization became a matter of life and death. Whalemen equipped themselves for a variety of emergencies: they kept a hatchet handy so that they could quickly cut a line if it tangled in the boat or, worse, around the limb of a sailor. They readied a bucket of water in the boat's stern to wet the taut, smoking whaleline as it was "braked" by the wooden loggerhead. They also took along an emergency kit. The kit on one ship contained a "water keg, holding three gallons, a lantern keg in which is a glass lantern, some bread, and tobacco & pipes, so that if the whale should run them out of sight of the ship they should have enough to live on for 4 or 5 days." "Each boat," the diarist explained, "is provided with a horn to use in case of a fog or night comeing on. They have also small flags to stick in the whale so that he can be seen from the ship if they should leave him to chase others, and a small compass if they should [lose] the ship and have to steer for the land."[8]

Before boats could be launched or whales could be killed, though, the prey had to be sighted. Novices tried to learn quickly the difference between a shark and a whale, and between a whale's spout and a whitecap, and they practiced repeatedly the difficult task of staying alert at the mastheads. For two hours at a time, they stood duty at the head of the topgallant masts. Perched on what Herman Melville called "two thin parallel sticks" (also known as the "t'gallant cross-trees"), they visually swept the sea for signs of a spout. Melville claimed

---

[6] *Port*, or *larboard*, is nautical language for *left*, and *starboard* means right. The larboard side of the vessel was the left side facing forward. Bark-rigged vessels generally carried five whaleboats – two on the starboard side – with space between them to cut in the whale. See Bockstoce, *Whales, Ice, and Men: The History of Whaling in the Western Arctic* (Seattle: University of Washington Press, 1986), p. 40.

[7] Charles M. Scammon, *The Marine Mammals of the Northwestern Coast of North America Together with an account of the American Whale-Fishery* (1874; reprint, New York: Dover, 1968), pp. 222–26; Hohman, *American Whaleman*, pp. 156–8; Bockstoce (*Whales, Ice, and Men*, p. 41) tells us that in the bowhead fishery the amount of line carried was sometimes 400 fathoms, reflecting the propensity of bowheads to run under the ice.

[8] Bockstoce, *Whales, Ice, and Men*, pp. 40–1; Diary of William A. Allen, *Samuel Robertson*, 26 September 1842, Log 1040, WMLOD.

that the masthead was a place for imaginative reverie, but he also acknowledged that a greenhand felt about as "cosy" at the masthead as he would "standing on a bull's horns." Sailor Robert Weir concurred. He was seasick after his first stint at the masthead, where he had been sent with the orders to look for "whales and whatnots." He saw two whatnots – two sharks – and his first impulse was "to throw myself to them for food."[9]

As greenhands made the rough adjustments to life and work at sea, experienced mariners were alternately exasperated and amused. The sailor who sighted a "large scool of flying porpoises," and the man who shouted at his shipmates, when he saw his first whale, that they should "see that smoke come out from the top of his Head" were fodder for veterans' journals if not for public ridicule. Captain A. G. Goodwin could barely contain his hilarity over the ignorance of a novice on his ship. After completing an assignment of mending boat sails, this sailor approached Captain Goodwin for assistance. "He came to me very innocently," noted the shipmaster, "& says where shall I make the button holes ha! ha! Only think of buttoning on a sail to the mast with horn buttons."[10]

Captain Goodwin enjoyed a good chuckle at his sailor's expense. Confronted with similar sorts of inexperience, other masters were less lighthearted. "I do not know what I shall Do," remarked Captain Stephen Stilwell on the *Rosalie*. "I have been sent [off] on A voyage with orders not to return again unless with A full ship. I have not a man before the mast that ever was a whaling Before." Furthermore, he said, there are "not more than 2 Before the mast knows his compass or knows how to steer." Stilwell was made painfully aware of the latter situation when he attempted to chase blackfish one day. He went in a whaleboat after the animals but then made the startling discovery that "there was not one on Board [of the *Rosalie*] that knew how to Back the mainyard." His solution was "to give up the Black fish & chase the Ship."[11]

A greenhand's inexperience did not usually last long into the voyage. If the ridicule of a man's shipmates did not make him a quick study, then the perils of whaling work fast underscored the necessity of knowing the business. Shipmasters, too, were not hesitant about using their disciplinary power to train new sailors. The captain of the *Roman* sent the entire starboard watch of that vessel aloft one night in September 1854 because they had not memorized the watch schedule. Faced with a "miserable set of green hands," Captain Charles Starbuck of the *Peru* was "determined to kill or cure." His method, noted a boatsteerer, was "a good one it is this he drew a map of a compass and gave it

[9] Herman Melville, *Moby Dick* (1851; reprint: New York: Norton, 1967), pp. 136–7; Diary of Robert Weir, *Clara Bell*, 22 August 1855, MSM.

[10] Diary of Robert Nathaniel Hughson, *Java II*, 3 February 1858, PC; Diary of George Blanchard, *Pantheon*, 22 December 1842, PC–TB; A. G. Goodwin to Swift and Allen, 20 December 1844.

[11] Diary of Stephen Stilwell, *Rosalie*, 15 August 1835; 26 August 1835, Log 1085A, WMLOD.

to two of them to learn and stopd there watch below until they learn it and keeps them to the wheel through the day he will fix them I guess."[12]

Faced with threats from all fronts, green whalemen soon got to know their way around their working home. Then they waited for what they were sure would be the culminating event of the voyage: the whale chase. They knew enough of the reputed hazards of whale killing, even before they saw their first spout, to anticipate the event with trepidation and awe. Some of them had seen the Seamen's Bethel back in New Bedford, where marble tablets commemorating whalemen killed in action lined the walls of the chapel. Herman Melville's whaleman Ishmael, who had read those tablets on a "darkened, doleful day," claimed that few sailors in that port could miss them.[13] Perhaps they feared what would come because they remembered too well the theatrical action pictured on whaling broadsides. Posters that promised would-be whalemen thrilling adventures on the high seas also warned them of impending disaster. Experienced shipmates, who readily shared tales of horror and woe, did little to calm the nerves of new whalemen. Seaman Samuel Chase expressed the feelings of many novices as he anticipated his first whale: "[I am] liable," he said, "in prosicuting our voyage to have not only my brains but my whole entire system dashed to atoms."[14]

Four diarists convey the character of the hunt, from its drama to its anticlimax, in especially evocative detail. William Alfred Allen, from Westbrook, Maine, was twenty-one years old when he signed aboard the *Samuel Robertson* in 1841 as a foremast hand. He was bound "Round Cape Horn to Cruise for Sperm Whales." Allen had long dreamed of his encounter with the cachalot, and he wrote down his experiences for his younger brother, who had similar ambitions.[15]

Fourteen years later, Robert Weir of West Point, New York, who left home after "wronging" his father, set out at age twenty to search for right and sperm whales. Unlike most deepwater whalemen, who headed for grounds in more distant oceans, Weir and his shipmates whaled in the waters of the south Atlantic.[16] They did not have good luck. By November 1857, Weir had endured month after month of barren seas. "Every day the same dull tune is played," he said. "Oh for the exciting Cry – there blows – to raise an excitement – to lower & chase whales even if we do not succeed in getting any – would relieve the monotony of this abominable life on board ship."[17]

William Abbe faced the same waste of a whaleless ocean in March 1859. He had gone to sea to recover his health. Not only did shipboard fare seem to

---

[12] Diary of James Cather, *Roman*, 22 September 1854, KWM; Diary of anonymous boatsteerer, *Peru*, 5 June 1851; 29 May 1851; 8 June 1851, NHA.   [13] Melville, *Moby Dick*, p. 41.

[14] Diary of Samuel Chase, *Arab*, 21 November 1842, MeHS.

[15] Diary of William A. Allen, *Samuel Robertson*, 23 October 1841, Log 1040, WMLOD.

[16] Weir diary, 19 August 1855.   [17] Ibid., 15 November 1857.

First Fort Dauphin Whale.

Robert Weir was twenty years old when he left New Bedford on the bark *Clara Bell* in 1855. From a prominent West Point, New York, family, Weir went to sea after "wronging" his father. He took his sketchbook with him, and his depictions of the whale hunt and of shipboard life and labor are some of the most detailed scenes that survive. In this sketch, a whaleboat is about to be crunched or, in the vernacular of the day, stove by the teeth of a sperm whale.

endanger his well-being instead of strengthen it, but the whaling was poor as well. He and his mates before the mast contemplated the continuously empty ocean and, in the spring of 1859, planned to desert ship at the first opportunity.[18]

Henry Howland did not have that freedom. As first mate of the *Good Return*, he felt bound to his ship, whatever its luck. And now, in the fall of 1830, he seemed to be having very little. "Oh merciful God," he wrote, "teach us resignation."[19]

"There blows!" was the shout that would eventually startle all these sailors out of their dull somnolence. Sung out from the masthead, it would, in an instant, galvanize seamen into frenzied action. Then new sailors would be thrown into work that was more grueling, more thrilling, and, in its denouement, more greasy than any they had known before. Then they would find out firsthand what veteran sailors meant when they said that whaling was hazardous work. In fact, for one of these four men – William Allen, Henry Howland, William Abbe, and Robert Weir – it would be a gamble with death that did not pay.

It would not be William Allen. Allen met up with several sperm whales, but

[18] Abbe diary, 12 March 1859.
[19] Log of Henry Howland, *Good Return*, 4 September 1830, Log 743, WMLOD.

he survived them all. He emerged from the experience, though, in complete awe of the challenges of the chase and the kill. He equated the actual attack on the sperm whale with "pricking the hind legs of an ugly horse."[20]

Things began, explained Allen, in a safe enough fashion. After sailors determined the position and direction of whales, the helmsman maneuvered the ship to come to windward of the animals. At the same time the crew scrambled to their whaleboats, and boatsteerers cleared their craft to be ready to lower. Then when the captain thought it was time to go, he gave the word. "Away go the boats," Allen said, "hand over fist the first down the best fellow and then for the struggle."

As the oarsmen strained their muscles to shoot the boat through the water, mates and boatsteerers hurried them on: "Pull ahead! do pull!" shouted the mate in Allen's boat. "I'll give you my hat! You shall have my chest of clothes! pull you Buggars!"[21] Boats approached whales either dead ahead or directly from behind to avoid being seen. Sometimes, claimed Allen, they sailed directly onto the whale's back. Whichever way they "go on" to the animal, though, said the sailor, they were in for a fight:

> If two boats get near the whale at the same time one of them heaves up, while the other goes on. The whales are generally going about 3 miles an hour and consequently the boats crew are some time within hearing of the unconscious creatures powerful breathing. Bye and Bye the mate sings out to the boatsteerer to stand up! and then the crew know that the moment is soon to come when their fate will be decided, whether they are to be killed or maimed, to have their boat stove or not.
>
> When they are in a proper position the mate sings out give it to him strong and then the almost fearful cry of stern all! Stern for your life everyman while the boat is dancing like a buble in the froth and foam. In an instant and almost before a man can grasp his oar you see his broad and quivering flukes or tail rising high in the air and wo to the boatscrew whose boatsteerer is not gifted with presence of mind enough to throw the stern of his boat out of the way for with such swiftness that you cannot trace it down it comes with a crash that can be heard for miles around.
>
> Now a long breath is drawn and the line which has been suffered to run out is now checked and a turn taken round the loggerhead, and away goes boat and whale . . . and if it is anyways rough the way the waters fly over her is a Sin! Trim boat says the mate Trim boat say the men and all they have to do is to hold on and go it. By and by the whale begins to slack his pace . . . The boat comes up to him and the mate throws his lance into

[20] William Allen diary, 26 September 1842. Unless noted differently, all quotations on the subject of the whale chase and killing in subsequent paragraphs are from this source and date.   [21] William Allen diary, 12 November 1841.

Three harpooneers stand ready to deliver a dangerous thrust, but their prey has "turned flukes" and gone into a deep dive. The whale's disappearance brings forth a lament from sailor/artist Robert Weir.

him. Hold on to him says the mate to his Bowman who has hold of the line and the mate keeps churning the lance into him.[22]

There he lies. Apparently he is dying and there is nothing to the eye of the inexperienced more to be feared but sudenly a strong trembling of his hump and quivering of his fins and flukes warns the mate of what is to come. Stern All! Stern every man! and in an instant the mighty creature a moment before so quiet is now in his flurry. With a giant power he throws himself almost clear of the ocean . . . and snapping his tremendous jaw with inconceivable rapidity. Now he rushes towards the boat and it seems as though nothing could save her from ruin when in a moment he turns and the danger is past and he is fin out or dead.

The first time Robert Weir's ship, the *Clara Bell*, took a whale was the first time it lost a boat's crew. When sailors aboard the *Clara Bell* spotted a whaleboat with a right whale on the morning of November 23, 1855, they also noticed that the starboard boat was gone. The captain ordered all hands aloft to search the seas. It was not long before sailors spotted a piece of an oar, floating. There was "scarcely a breath heard from our mouths," said Weir. They then sighted a man's hat. The crew was "sick at heart." By 1:00 P.M., when there was still no sign of the missing whalemen, the feeling of the men of the *Clara Bell* "may well be imagined." But at 2:00 P.M., a boatsteerer thought he "saw something" off the ship's weather beam. The captain leveled his spyglass on the "speck." He wheeled about and cried out to his sailors, "Get you dinner boys and breakfast too – theres the boat alongside of the dead whale!"[23]

[22] The officer who pumped the lance up and down into the whale hoped to strike the whale's "life" (its heart or lungs). This mortal blow made the whale produce the telltale sign of a lethal strike: a spout of red blood. Bockstoce, *Whales, Ice, and Men*, p. 42.

[23] Weir diary, 23 November 1855.

The sight of that whale, and the other "mighty monsters" that Weir killed (and survived) in his voyage from 1855 to 1858 repaid him "for all of the trials [he] had . . . gone through." That was fortunate, for Weir's voyage made little in the way of money.[24]

William Abbe survived his voyage, too, possibly because he did not come to know whales very well at all. Although he encountered enough of the hunt and of trying out to tell him he did not need to repeat the experience, his voyage was, like that of Robert Weir, a failure. A full ship was a vessel with 1,000 barrels of oil. The *Atkins Adams*, for its four-and-a-half years' effort, sent home only 275 barrels of sperm oil.[25]

Henry Howland had far better luck seeing whales and far worse luck when it came to outliving them. Within three days after he lamented his poor whaling luck in December 1830, his ship began taking good-sized right whales, each yielding up to 150 barrels of oil. In the early morning of January 15, 1831, he was busy cleaning the baleen from some of this catch when the man at the lookout spotted some sperm whales. Howland quickly gathered his boat's crew together, and they lowered the starboard boat. Another whaleboat succeeded in harpooning a whale, and Howland's boat attached itself to the same animal. But then the whale turned and, throwing the weight of its head on Howland's boat, stove it. At the same time, it closed its jaws around Howland himself and carried him under water. Two minutes later the mate rose next to the whale, but "before assistance could reach him, he sunk and was seen no more." The whale then turned upon another boat, caught a sailor named Simeon Hersey in its mouth, "hurt him considerable," but released him alive. Finally, after spouting blood, the whale swam away. "Our feelings," noted the man who took over the job as first mate, "can easier be guessed than described."[26]

Disaster and drama were real and significant parts of whaling but largely because the threat of tragic events loomed large over a voyage. Deaths like that of Howland were not actually common in these accounts.[27] And fights with sperm

---

[24] Weir diary, 23 November 1855; 9 December 1855; 25 May 1856. The results of the *Clara Bell's* voyage are listed in Alexander Starbuck, *History of the American Whale Fishery* (1877; reprint, Secaucus, NJ: Castle Books, 1989), p. 529.

[25] Starbuck, *American Whale Fishery*, p. 561.

[26] The Howland incident is described in detail in "The [New Bedford] Morning Mercury," 4 April 1925, cols. 5 and 6.

[27] General statistics on mortality in the whale fishery are hard to come by. Approximately one man (out of a crew of around thirty) died aboard these vessels during an average three-and-a-half-year voyage. About half of the deaths were the results of accidents; the others were caused by disease. More men among these crews may have died in port, of course. Elmo Hohman surveyed figures associated with fifteen whaling voyages (average length of voyage, 3.3 years) during the mid-nineteenth century and found 3.3 percent of the original crews were reported to have died; nearly 40 percent were discharged. See Hohman, *American Whaleman*, p. 316.

whales should not obscure the actual variety involved in deepwater hunting. Right whaling, for instance, could be considerably less challenging than sperm whaling. Even though the right whale's wide flukes could be dangerous, its lack of teeth made it less destructive and it was not as aggressively self-protective as the sperm whale. And bowhead hunting presented whalemen with a very different set of challenges. Bowheads were more sluggish and more easily attacked than sperm whales, and they were less ferocious even than right whales. But they often dove deep when they were harpooned, and when they surfaced they frequently headed for nearby ice. And men who chased bowheads contended with more than their prey. Their whaleboats were not infrequently swallowed by Arctic fog, their ships surrounded by ice, and, as any northern sailor knew, the cold water could put a fast end to any man who fell into the sea.[28]

The dangers of whaling, then, were diverse. The work also had an anticlimactic side. Thoroughly unexciting, but just as necessary to whaling labor, was the onerous job of cutting in and trying out the animal, and in the case of baleen whales, of cleaning and drying out whalebone. Well aware that stripping and boiling blubber did not have the romantic cachet of the hunt, whalemen nevertheless considered this an important aspect of their business, if only for its special miseries.

Sailors launched into the greasy, smoky, smelly task of processing the whale as soon as possible after it was brought alongside. Fearing that if they did not cut in soon enough, sharks might consume the carcass or wind might make the burdened ship unstable, they tried to work quickly. *Quickly*, however, was a relative word in this business, for cutting up a whale's body and trying out could take several days.[29]

Whalemen first secured the huge animal alongside the vessel with block and tackle, and then, standing on a "cutting stage" – a wooden platform that was positioned above the whale's body – officers severed the whale's head and began to cut the whale's blubber from its body. As if they were separating the peel from a large revolving fruit, they stripped the blubber from the carcass, turning the whale in the water as they did so. Up on deck, the foremast crew worked at the windlass to heave the long, heavy strips of blubber, known as "blanket pieces," on to the deck. The blanket pieces were sent immediately down into the "blubber room," where boatsteerers cut up the huge strips into smaller "horse pieces." Meanwhile, sailors worked on the severed head. They stripped baleen from the jaw of baleen whales and took case oil and spermaceti from the huge head of sperm whales. They used long ladles to dip out the case oil, or

[28] John Bockstoce, *Whales, Ice, and Men*, pp. 66–74.
[29] Hohman, *American Whaleman*, pp. 167–170.

Cutting in

As sharks hungrily cruise in the bloody water below, sailors on the cutting
stage strip away and haul aboard the thick, heavy blubber of a dead whale.
(Robert Weir sketch.)

they sometimes sank into the whale's head themselves to retrieve as much of the
precious substance as they could.[30]

And then came the hot, smoky part – that aspect of whaling that was,
according to seamen, nothing short of ghastly. The tryworks were fired up, and
the horse pieces, minced into smaller sections about half an inch thick, were put
into the trypots. The oil that separated from the boiling blubber was ladled into
copper containers for cooling and eventual storing, and the remaining material,
known as "cracklings" or "fritters," served to fuel the fire further. The flames
from the trypots, the smoke from the fire, the smell of the boiling blubber,
along with the greasy, slimy decks, made for a situation that seamen described
as truly hellish.[31]

"It would be a curious sight to a stranger," commented William Allen, "to
come on board of a ship trying out in the night. The fire from the try works
shooting up half as high as the main top illuminating the large sails [made] the

[30] Ibid., pp. 171–3.
[31] Haviland diary, 22 June 1857; William Allen diary, 26 September 1842; Hohman, *American
Whaleman*, pp. 171–4.

In this sketch, sailors, one naked at least to the waist, dip deep into the huge head, or "case," of a sperm whale to tap its reservoir of clear oil. (Robert Weir sketch.)

The sperm whale was prized for more than its oil. Here sailors from the *Clara Bell* in 1855 pull out a line of teeth from the whale's lower jaw. There was considerable competition for the best and biggest teeth, which with ink and a sharp point, a steady hand, and a modicum of skill could be transformed into a piece of scrimshaw. (Robert Weir sketch.)

whole ship look as though she was on fire, while the men could be seen flitting about just as greasy all over as though they had been dipped in oil and their faces looking as shining as so many looking glasses." Robert Weir had the same feeling. The sight of men trying out whale blubber was such an unearthly one that he would have liked his friends on shore to see him at it: "Wouldnt it be gay if the folks at home could see us now – up to our knees in gurry &

Robert Weir depicts one of the most arduous aspects of whaling labor: trying out. The blubber is cut into small pieces (one is being lifted into the trypots here) and boiled down (or "tryed out"). The oil that results is cooled and then put into large casks (witness the man at left) for storage. The scraps of the whale's skin that remain are recycled as fuel for the fire.

grease . . . the men who work in the waist almost suffocating they look like so many devils."[32]

For disgusted expressiveness on the subject of trying out, though, no one can come close to William Abbe. On July 17, 1859, Abbe and his shipmates boiled blubber for eighteen hours straight. His clothes were soaked in oil; his body was "covered from crown to heel with oil." He ate "with oily hands oily grub" and drank from "oily pots." It is not hard to imagine what Abbe dreamt of when he was finally sent below to get some sleep: "[You] dream you are under piles of blubber that are heaping & falling upon you till you wake up with a suffocating sense of fear and agony only to hear the eternal clank of the cutting machine & the roar of the fires under the try works – or the wind dismally howling through the rigging – to fall asleep only to dream again till you are called on deck." Consciousness, of course, did not improve the situation. Abbe's eyes smarted with pain from the black smoke; his face and hands were burned from the "sputters and spatters" of boiling oil; he was "weary – dirty – oily – sleepy – sick – disgusted." And there he sat on the horns of a dilemma: "To go through such a scene [again] – I confess the very thought turns my stomach & dizzies my head – yet I am not foolish enough to wish this valuable ship to be idle."[33]

[32] William Allen diary, 26 September 1842; Weir diary, 24 July 1857.
[33] Abbe diary, 17 July 1859.

Trying out whales was universally arduous, but it was not done identically in all branches of the fishery. The right whale and the bowhead whale did not carry a built-in reservoir of oil as did the sperm whale, so all the blubber of these whales had to be tried out. And then there was the baleen to deal with. Whalemen sometimes extracted 2,000 pounds of it from a single right whale, and 3,000 pounds from a bowhead. They then had to split the baleen, scrape it clean of all "gum," and then wash, polish, and dry it. Moreover, these whalemen could not cut corners. Baleen that was not carefully cleaned sometimes smelled, and baleen that was not carefully dried tended to shrink.[34]

Whaling, like any work tied to the vicissitudes of nature, had its periods of intensity and its slow moments. With a large crew to share sailing work and with no place special to go, sailors found that cruising without whales around could be downright relaxing. William Allen had "first rate times" on the offshore grounds in the eastern Pacific. Another man claimed that his job meant little more than "to eat help make sail and take it in." It "almost seems as if I was on a visit," he declared. Another described his "only real work" as breaking out water and beef casks. A fourth man enjoyed his ship's regular "jollifications" and a fifth, who cruised in the Arctic, did no work "except that what is absolutely necessary" and had "gay times in the forecastle singing playing on fiddle flute tamborine & bones & living on Duff everyday a sailors greatest dish."[35]

Sailor Edwin Pulver was so impressed by the leisurely pace of life on his ship that he wrote a poem in honor of his less than rigorous routine:

> Verry good times we have indeed
> Nothing to do but sleep and feed
> But sleep is the best I think of the two
> For theres No Exertion to go through
> But in Eating we have to use our paws
> To pass the grub into our jaws
> And when A man Eats a great deal
> It will make him rather tired feel.[36]

Days of doing little but sleeping, feeding, and enjoying "jollifications" made some men happy, but most mariners looked at quiet times as a mixed blessing. Sailors who had time to socialize were obviously men who were not busy whaling, and "Easy Times," as sailors put it, were also dull times and hard times. Seamen complained, in fact, that days on an idle ship could merge into a seamless round of depressing monotony. A sailor aboard the *Canada* in 1847 offers something of the character of life in a whaleship on a whaleless sea:

[34] Hohman, *American Whaleman*, pp. 180–2; Bockstoce, *Whales, Ice, and Men*, pp. 74, 82–3, 95.
[35] William Allen diary, 27 May 1842; Diary of James Allen, *Alfred Gibbs*, 9 August 1870, PPL; Abbe diary, 21 August 1859; Martin diary, 20 April 1842; Haviland diary, 14 May 1857.
[36] Diary of Edwin Pulver, *Columbus*, 30 June 1852, PPL.

October 30: Sun went down, wind went down the Same time and left us
in a dead Calm. The Sails hanging from the yards. Lifeless. the Moon
Shone bright. all hands round the Deck Spining Yarns. What a time for
Sailors. Hay. November 2: . . . Nothing Seen and nothing done. bad luck
to the whals November 3 . . . somthing was done this day but i dont know
what it was now, anyhow it began at 7 A.M. and finished at 2 P.M what
it was i cant remember. nothing more was done and nothing more was seen
either. 16 November bent sail . . . Nothing farther was done and nothing
was Seen that is what we say Every day, nothing Seen what a word that is
that nothing i mean. When a Man comes down from the Mast Head, the
Captain Stands there well what have you seen this morning. Nothing Sir,
Confound that Nothing, what dose it look like i dont know Sir and so it
goes on from one day to another.[37]

"Day in day out," echoed another whaleman, "week in and week out year in
and year out, and then what? And the same thing again. Well what well nothing. Positively nothing excepting it may be called something to be ever viewing
sky and water Sky and water This way That way or Which way you may it is
sky and water."[38]

Hard work on a whaler, then, was usually linked to good hunting, and easy
times were tied to bad luck. The whaleman was inevitably a person of mixed
emotions, who simultaneously read into his situation satisfaction and discour-
agement, success and failure.

II

The sailors who spent the better part of three to four years on isolated seas,
contending with ennui or overwork, relaxed contentment or desperate exhaus-
tion, brought much more to their endeavors than their capable hands, willing
energy, and their sense of humor or outrage. They came whaling with ambition
and interest and endowed their experience with distinctive meaning. Most
seamen claimed that surviving the whale hunt enhanced their manhood. At the
same time, however, they recognized that whaling was a social liability. Victorian outsiders, for their part, sought and found distinctively genteel significance
in the deepwater experience.

Whalemen who came to sea from working-class backgrounds, as many did,
valued whaling for the way it challenged and hardened them as men. Like many
of their shore-based counterparts, these sailors wanted to be tough and physically
intrepid.[39] Although they did not dismiss the pleasures of peaceful life and

---

[37] Diary of Sylvanus C. Tallman, *Canada*, 30 October, and 2, 3, 16 November 1847, Log 200,
WMLOD.   [38] Diary of Ambrose Bates, *Nimrod*, 18 March 1861, KWM.
[39] On expressions of working-class masculinity, see Elliot J. Gorn, *The Manly Art: Bare-
Knuckle Prize Fighting in America* (Ithaca: Cornell University Press, 1986), pp. 140-1; Sean
Wilentz, *Chants Democratic: New York City and the Rise of the American Working Class,
1788-1850* (New York: Oxford University Press, 1984), pp. 256-65.

leisure, they took pride in fighting ocean storms. And when they battled large
whales – "monsters of the deep," as they frequently called them – they certified
their bravery.

Cape Horn, with its headwinds, crosscurrents, violent weather, and forbid-
ding aspect, was the prime testing ground for nautical daring. Experienced
mariners tried to heighten the rigors of rounding the Horn as best they could.
Veterans aboard the whaler *Alfred Gibbs*, for example, enjoyed frightening
greenhands about the horrors of the Horn. Greenhand James Allen reported
that "old sailors" aboard that ship sought to terrify neophytes with "what a
dangerous place it is &c each one . . . trying to tell bigger lies than the others."
When they passed a trial like Cape Horn, mariners liked to compare their
daring with the alleged timorousness of landsmen. One whaleman who endured
the Horn put his sentiments on this subject into verse:

> [The wind] increased at night, untill it blew a gale
> And though twas not much to a sailor's mind
> A landsman would have looked a little pale
> For sailors are in fact a different kind.[40]

It was the fight with the leviathan, and particularly the sperm whale, though,
that entailed the most daunting contests. Alexander Hoxie's meeting with a
sperm whale in early March 1833 reveals the grand accomplishment that a man
might feel as a veteran of a whaling battle. For the first time in his life, Hoxie
explained, he was "fighting with and killing a large whale." This "monster," he
boasted, made "the deep boil like a pot." Furthermore, whales like this beast
were so strong that "one attacked a Nantucket ship about this latitude a few
years since and sunk her." . . . "I was previously fearful my courage would fail
me when coming onto the whale," announced Hoxie in retrospect, "but I have
enjoyed the scene very highly and have established my courage in my own mind
at least."[41]

Hoxie bolstered his bravery by alluding to sperm whale ferociousness and by
referring to the demise of the whaleship *Essex*. In November 1820, the *Essex*,
which had been cruising in the mid-Pacific, was rammed twice by a sperm
whale. The ship sank quickly, and twelve out of the crew of twenty were lost in
whaleboats or died of exposure.[42]

After 1850, the *Essex*'s whale was not the only powerful animal a sailor might

[40] James Allen diary, 20 November 1870; Blanchard diary, *Pantheon*, 4 February 1843,
PC–TB.
[41] Diary of Alexander Hoxie, *Maria*, 2 March 1833, quoted in Edouard A. Stackpole, *The Sea-
Hunters: The New England Whalemen During Two Centuries 1635–1835* (Philadelphia:
Lippincott, 1963), p. 407.
[42] Stackpole, *Sea-Hunters*, pp. 317–37. Other accounts of sinking of whaleships and attacks on
whaleboats can be found in Stackpole, pp. 334–5, and in Starbuck, *American Whale Fishery*,
pp. 116–29.

use to glorify his fight with a sea monster. Benjamin Boodry, mate of the *Arnolda* of New Bedford, had been reading *Moby Dick*, and he knew that Melville's whale, which was endowed with unearthly power and meaning, could be an emblem of his own importance. In the winter of 1853, Boodry encountered a mean sperm whale. This animal stove two whaleboats, threw Boodry into the water, and then swam away with five harpoons in its flesh. "This being knocked 15 feet into the air and comeing down alongside of his jaw by a whales rooting is not what it is cracked up to be," he remarked with bitter irony. It seems, though, that Boodry's misery was diminished by the grand nature of his antagonist. It was, he said, a meeting with "my evil genius," a contest with none other than "Moby Dick."[43]

Whalemen not only drew on historical sources and literary works to glorify their prey (and themselves), they also turned to whaling songs. "The noble whale," explained George Blanchard in 1847, was "the game worth killing & which makes a man proud to look upon, to feel that he can conquer the King of the Seas." Blanchard underscored his feelings with the lyrics of a sea song:

> Oh! the rare old whale 'mid storm and Gale,
> He lives for a thousand years,
> He sinks to rest, on the billows breast,
> Nor the roughest tempest fears.
>
> The howling blast, as it rushes past,
> Is music to lull him to sleep,
> And he scatters the spray, in his boisterous play
> As he dashes – the King of the Deep.[44]

George Blanchard and others like him may well have won their battle for manly self-respect. Or they may not have. The whale and the ocean, it seems, were not the only antagonists, or even the most formidable antagonists, these men encountered. Challenging these whalemen at every turn, on shore and at sea, and threatening to undermine any manly authority they gained, seemed to be individuals armed with some of the most powerful weaponry available: social scorn. Men who went whaling, George Blanchard declared, were looked upon by "judging" people in America with "contempt."[45]

Blanchard's lament echoes throughout the journals of whalemen. The far Pacific and the frigid Arctic apparently meant no escape from domestic clashes of culture and class, and sailors felt the far-reaching effects of social prejudice.

---

[43] Diary of Benjamin Boodry, *Arnolda*, 10–11 January 1853, Log 619A, WMLOD. Melville's novel, which was published in New York in 1851 to mixed reviews, was not widely read by either whalemen or landsmen. Boodry's familiarity with the novelist and his notorious whale was thus unusual.

[44] Blanchard diary, *Solomon Saltus*, 9 August 1847; Herman Melville concluded the "Extracts" section of *Moby Dick* (p. 11) with a version of this song.

[45] Blanchard diary, *Solomon Saltus*, 19 August 1847.

And they shouldered the burdens of condescension from all corners, even from those whom they might have expected to be the most loyal: whaling women. "I heard the mate remark the other day," noted steward Silliman Ives in 1868,

> that he knew of a young lady in Nantucket who refused to dance with filthy whalemen. Incredible preposterous in the extreme, an idea not to be entertained for a moment. Oh, Nantucket mother of whaling! and has it come to this that thy fair daughters – they who erst did delight to honor the hardy sons of Neptune who go down upon the mighty deep and beard the leviathan on his own cruising ground and convert him into gold, or more correctly speaking greenbacks, can it be possible that in these latter days, thy fair women have soured on these brave men? Have come back on the bold mariner? No. I'll not believe it.[46]

But Ives had better believe it, for the attitudes expressed by the "daughters" of whaling were, as whalemen knew, widely shared.[47]

It is not clear whether whalemen had only recently been denigrated as social pariahs. Prejudice against sailors in general was not new to the 1800s. In the seventeenth and eighteenth centuries, Puritans worried about men who worked outside the family and apart from the constant governing eye of employers. The key to social order, they felt, was continuous, disciplined work within a household setting. Men with irregular work habits who were forced to roam about were objects of disapproval.[48] Early whalemen may have faced additional prejudices as well. Whaling was in essence a hunt, and hunting, in the minds of those acquainted with "modern" hierarchies of human progress, was primitive. Add to this the fact that hunters par excellence – Indians – made up a major part of the labor force in the Colonial period and the whale fishery became further suspect.[49]

Whalemen of the nineteenth century may have tapped into longstanding biases, but they may also have suffered from new social standards. They were seen as adrift from the institution that increasingly stood at the center of moral and social redemption: the home. With the advent of longer and longer voyages, whalemen were considered unreceptive to the "civilizing" influence of domestic circles.[50] Whaling vessels also accommodated men of different races. Although

---

[46] Diary of Silliman B. Ives (under pseudonym of Murphy McGuire), *Sunbeam*, 27 September 1868, Log 618, WMLOD.

[47] See, for example, remarks in Pulver diary, 12 December 1851; and in *The First Annual Report of the American Seaman's Friend Society* (New York: J. Seymour, 1829), p. 28.

[48] Daniel F. Vickers, "Maritime Labor in Colonial Massachusetts: A Case Study of the Essex County Cod Fishery and the Whaling Industry of Nantucket, 1630–1775" (Ph.D. diss., Princeton University, 1981), pp. 10–20.

[49] Native Americans were not invariably hunters, but they were treated as such and also treated as primitives by politicians and intellectuals. See Reginald Horsman, *Race and Manifest Destiny: The Origins of American Racial Anglo-Saxonism* (Cambridge, MA: Harvard University Press, 1981), pp. 107–8.

[50] The literature of the New Bedford Port Society and the American Seaman's Friend Society, established respectively in 1825 and 1830, is replete with such implications.

the labor force of the fishery had always been racially integrated, it more and more depended on seamen of color from distantly "primitive," even cannibalistic, areas. To a white society drawing lines between races ever more distinctly, and flourishing new "scientific" proof to underscore its supremacy, the fishery may have appeared to flout important proprieties.[51]

Not least, men in the whale fishery may have suffered because they were newly seen as unclean. Nantucket women, according to the rumor just re-counted, snubbed whalemen for their "filthiness." Americans' commitment to cleanliness apparently "grew ever stronger" throughout the nineteenth century as a means of asserting social position and as a means of social mobility. "Dirty hands, greasy clothes, offensive odors, grime on the skin," two historians argue, "all entered into complex judgements about the social position of the dirty person and actually about his or her moral worth. By the middle of the nine-teenth century," they claim, "personal cleanliness ranked as a mark of moral superiority and dirtiness as a sign of degradation."[52]

It is not hard to imagine how whalemen fared in the hierarchy of cleanliness. When whalers were engaged in the business of taking or trying out whales, they challenged every standard of Victorian hygiene. Men's faces, as we have seen, shone like mirrors with grease; their clothes smelled of urine, smoke, and blubber; their skin was caked with grime. The ship's deck itself was deep in oil, blubber, and blood – "gurry," as sailors called it. Even though whalemen cleaned themselves and their ships before sailing home, merchant seamen and navy crews were witnesses to their at-sea appearance and carried their unfavorable impressions back to the United States. One whaleman summed up the looks of whaling work, and by association, the industry's reputation in genteel company: "Everything," he said, is "beshit."[53]

Whalemen, then, did not stand well in the social strata of industrializing America. Their work was too erratic, their hands too soiled, their company too mixed, and their ships too far from home to be acceptable to the middle-class arbiters of status. Whalemen were not the only targets of domestic social con-cern, of course. Among others, men and women who took jobs in mills and manufactories and who carried with them "uncouth" rural behaviors, or those who were "idle" or "dissolute" immigrants, or the many individuals who did manual labor were similarly denigrated. But in a changing America, whalemen seemed to represent the very worst qualities of an old order.[54]

[51] George M. Fredrickson, *The Arrogance of Race: Historical Perspectives on Slavery, Racism, and Social Inequality* (Middletown, CT: Wesleyan University Press, 1988), pp. 201–5.

[52] Richard L. Bushman and Claudia L. Bushman, "The Early History of Cleanliness in America," *Journal of American History* 74 (March 1988): 1213–38.

[53] Diary of William W. Taylor, *South Carolina*, 21 November 1836, KWM.

[54] Edward P. Thompson, "Time, Work-Discipline, and Industrial Capitalism," *Past and Present* 38 (1967): 56–97; and Herbert G. Gutman, *Work, Culture & Society in Industrializing America* (New York: Vintage Books, 1977), pp. 3–78, pointed out some time ago the

Whalemen felt slighted by the shoreside middle class, but it was not simply on land that they faced contempt. Merchant sailors at sea, whalemen found, could be especially arrogant, and meetings with trading ships were often occasions for social snubbing. Men on the *Mount Wollaston* spotted a merchant ship at 7:00 A.M. on July 20, 1854: "[We] braced round and stood for [her] but as she would not speak us, we took in sail and laid the main yard aback." The schooner *S. R. Roper* was whaling in the south Atlantic in 1865 when they saw a merchant bark. Hoping to hear recent news and to get the correct latitude and longitude, the captain sent a boat's crew over to the passing vessel. "What do you want," the skipper of the bark asked the whalemen "in a gruff voice." He refused to invite the whalemen aboard – "very impolite indeed," noted the whaling diarist.[55]

In a final example, the captain of the whaleship *Arab* wanted to send letters aboard a nearby clipper in 1855. He worried, though, about how he would "conciliate the skipper of the merchantman if he should be angry that his meditations were interrupted by the captain of a 'blubber hunter.'" He finally decided to send over to the merchant ship a boat's crew bearing gifts of fresh vegetables as "a peace offering."[56]

Evidence from the diaries of merchant sailors helps to explain why merchantmen might have been so snobbish. According to merchantmen, who were skilled specifically in sailing and who devoted every scrap of energy to manipulating wind and water, the whaleman was not a true sailor. He was a literal "jack of all trades" who was handy, not expert, at numerous jobs. He managed to make a ship go, for sure, but he also rowed boats, hunted animals, and boiled blubber. And to look into a whaling forecastle was to see many rubes from the hills and haunts of inland America, not "modern" specialists in sailing. Richard Henry Dana expressed the attitudes of many merchant sailors when, aboard his brig *Pilgrim*, he encountered a whaleship in 1834. A whaleman who clambered aboard the *Pilgrim* was a "thoroughly countrified-looking fellow, [who] seemed to care very little about the vessel, rigging, or anything else, but went round looking at the livestock, and leaned over the pig-sty, and said he wished he was back again tending his father's pigs." Dana furthered the slight by asserting that

ongoing tensions in American industrial settings between men and women who embraced agrarian and artisan work habits and those supportive of an industrial ethos. Stuart Blumin has argued more recently that the reshaping of class lines in the nineteenth century was accompanied by a heightened distinction between nonmanual and manual labor, and the social denigration of the latter. See his *The Emergence of the Middle Class: Social Experience in the American City, 1760–1900* (Cambridge, UK: Cambridge University Press, 1989), p. 121.

[55] Diary of Charles Stedman, *Mt. Wollaston*, 20 July 1854, NBFPL; Diary of Caleb Hunt, *S. R. Roper*, 29 October 1865; 22 February, 26 February 1866, KWM.

[56] Diary of William M. Stetson, *Arab*, 27 October 1855, Log 507, WMLOD.

the whaler's skipper, whom he acknowledged was an experienced commander, "had not the slightest appearance of a sailor."[57]

The most haughty merchantmen were, not surprisingly, the officers of clipper ships. These men commanded those speedy, white-winged marvels that gained international applause in the mid-1800s and that represented the very best of the new business world: they made money through speed, competition, discipline, and punctuality.[58] The captain of one clipper commented proudly that on his ship, "everything [went] on like clockwork." Another remarked on the "mechanical regularity" of clipper ship labor.[59]

Officers of these merchant vessels, always in competition with the fastest passage on their particular route, were beset with worry if their own sense of urgency did not match the wind and weather. "Calm!" noted an exasperated first mate on a trading voyage to the East Indies. "Calm! Calm! . . . Calm!! calm!! C.A.L.M.!!" The master of another merchantman echoed him: "Calm Calm Calm Calm Calm Calm." When the wind did pick up, these eager mariners carried as much sail as they could and adjusted it to catch "every whiff and squall."[60]

Pressured as they were by owners to reach markets quickly, clipper captains had little time to chat with leisurely cruising whalemen. Isaac Baker, mate of a pepper trader making its way south in the Atlantic in 1855, noted that his captain would not even "signalize" with a nearby bark because he was not "near enough to the Equator yet" and the ship was making poor progress. Any delay, even to hoist signals, would be time lost.[61]

It was not simply that merchant seamen had no time to visit with whalemen; they also seem to have had little inclination. Focused as they were on the modern concerns of speed, they saw the whaler as an amusing and curious anachronism but not as a social equal. Isaac Baker passed whalers occasionally in the Atlantic and Indian oceans, and he made his sense of superiority quite clear. "At 3:30 PM a sail in sight at the Westd," he noted in December 1841. "At 10 passed her, she proved to be a 'spouter' jogging lazily along under topsails foresail & fore topmast staysail. . . . Spoke with the whaler ship no, no, *almost* spoke with her, not quite."[62]

---

[57] Richard Henry Dana, Jr., *Two Years Before the Mast*, ed. John Haskell Kemble (Los Angeles: 1840; reprint, Ward Ritchie Press, 1964), p. 34.

[58] On the rising clock orientation of Americans in the nineteenth century, see T. J. Jackson Lears, *No Place of Grace: Antimodernism and the Transformation of American Culture 1880–1920* (New York: Pantheon, 1981), p. 10.

[59] Diary of Edward Harrington [?], *Oregon*, May 1840, EI; Diary of Henry Davis, anonymous merchant vessel, 29 May 1862, MSM.

[60] Diary of Isaac Baker, *Merrimac*, 6 May 1858, PMS; Diary of D. F. Weekes, *Brothers*, 27 March 1866, PMS; Davis diary, 21 July 1862.

[61] Diary of Isaac Baker, *John Caskie*, 26 August 1855, PMS.

[62] Diary of Isaac Baker, *Taskar*, 27 December 1841, PMS.

Where Baker was haughty about the "retrogressive" ways of whalers, Captain Charles Emery was outraged. Emery, a merchant shipmaster, took passage home from Chile in 1839 in the whaleship *Columbus*. It was, he said, a journey through "purgatory." "Surely the bitter & grievous curse mentioned in the bible," he groaned, "must be taking passage home in a whaler." The problem centered, it seemed, on the slowness of the *Columbus*. Even during the daylight hours, with "moderate" winds, many of her sails were furled. It seemed clear to Emery that on this whaleship, there was "nobody but myself in a hurry."[63]

In the go-ahead business world of the mid-1800s, where Americans wanted to be on time, to be clean, to be efficient, and to work regularly, whalemen seemed out of date. Although they earned some credit from the whaling voyage by being brave, they incurred social censure by doing dirty and irregular work. The two features of whaling may not have been unrelated, of course. As whalemen faced heightened stigmatization among bourgeois circles for the work they did, they may have valorized the hunt, the chase, and the kill, knowing that courage and daring still had wide currency among men of their own kind.[64]

As they labored and wandered over foreign seas thousands of miles from home, Victorian whalemen, like their fellow sailors, felt the long reach of the American shore. They, too, carried to sea with them the imperatives of home society. In contrast to their shipmates, though, they sought the ocean world and the deepwater experience less as a route to masculine hardihood (although they did not eschew courage when the need for it came to them) than as a window on the sublime. At sea, nature's magnificence and God's omnipotence happily converged to lead them to new heights of experience. When Robert Weir beheld his first whale, he felt it a "blessing that I could behold a mighty work of God," and, likewise, when his ship struggled through a wild and frenzied ocean in 1856, he was thrilled and comforted by the scene. He tried to be on deck as much as he could, to "hear the Allmighties Voice in the storm." "It makes us feel," he said, "that we are actually in the great presence of the Omnipotent."[65]

---

[63] Diary of Charles Emery, *Columbus*, 11 May 1839; 23 June 1839; 17 April 1839, PMS.

[64] Elliott Gorn argues that white American workers in the nineteenth century, faced with failure to succeed as a man by Victorian standards, turned to "a more elemental concept of manhood . . . toughness, ferocity, prowess, honor." *The Manly Art*, p. 141. Donald Yacovone says that in the "Age of Jackson" the "frontiersman, Indian fighter, and mountain man became prototypes for the manly ethos," suggesting that the whaleman might eventually have exemplified even middle-class masculine ideals if he and the hand-to-hand technology that he used had survived longer. See Donald Yacovone, "Abolitionists and the 'Language of Fraternal Love,'" in Mark C. Carnes and Clyde Griffen, eds., *Meanings for Manhood: Constructions of Masculinity in Victorian America* (Chicago: University of Chicago Press, 1990), p. 85.

[65] Weir diary, 23 November 1855; 24 April 1856. Studies of Victorian manhood suggest that physical daring as a masculine ideal for members of the middle class may have taken hold

These sailors had a mixed reaction to the actual work aboard a whaler. Shipboard life, with its regimentation and disciplined routine, suited men committed to self-control. Whaling work itself, however, with its grease, its mess, and its unpredictability, was an unpleasant discovery. What was a man to do, though, who found himself on the far side of the globe from his genteel home, faced with work that he could not abide? The solution, not surprisingly, was the merchant service. William Abbe, foremast hand, was disenchanted with the "backward" ways of his whaler, and he met up one night with the ship of his dreams. A large clipper ship that had been astern of his whaler passed him at about two o'clock in the morning in the "crystal moonlight." The clipper had about twenty sails set; the whaler five. "I felt a longing to be aboard her," remarked Abbe, "and quit the 'old tub'. . . . Nothing more lovely than the night – nothing more beautiful than that ship – with her lofty sky sails & lower sails – her handsome hull & her eerie speed – I walked the deck impatient on my restraint on this laggard barke – & longed to be on a moving quick winged ship." "I have seen," he said," all there is to Whaling, & am curious to be among true sailors."[66]

Whaling, then, gave some men credentials in self-control and qualified others in courage and daring. It taught some sailors hard lessons in social prejudice but for others confirmed cultural biases. Whaling paid dividends, in part, according to a man's unique social investment and his singular set of expectations. But whaling taught some universal lessons, too. All men, whether they wanted to or not, and whether they sailed as Victorian visitors or were bred to the sea, learned to become part of a ship's company and a sailor's community.

Whaling linked men in distinct alliance in a multitude of ways. From the moment sailors suffered their first nausea at the start of the voyage to the time they tried their sea legs out in home port, they took part in a life that no landsmen shared. Whalemen came to know and use a singular technology and a distinctive language. They employed an esoteric nomenclature, and commanded a special terminology for the ship, the sails, the boats, the whales, and the work routine.

Whaling work also connected men to one another. The deep, cold ocean and the jaws and flukes of whales exacted fast, fatal punishment from men who did

only in the mid- to late nineteenth century (Theodore Roosevelt being the most prominent spokesperson of physical bravery). On this subject see John Higham, "The Reorientation of American Culture in the 1890s," in John Higham, ed., *Writing American History* (Bloomington: Indiana University Press, 1970); and Jeffrey P. Hantover, "The Boy Scouts and the Validation of Masculinity," in Elizabeth H. Pleck and Joseph H. Pleck, eds., *The American Man* (Englewood Cliffs: Prentice-Hall, 1980), pp. 285–301, and E. Anthony Rotundo, "Body and Soul: Changing Ideals of American Middle-Class Manhood, 1770–1920," *Journal of Social History* 16 (Summer 1983): 23–38.   [66] Abbe diary, 17 May 1859.

not work in close intimacy with shipmates. The watch system and the lay system forced men to put aside cultural compunctions and to collaborate. Both systems underscored the ways that the men of a whaleship – if we can forgive the expression here – were in the same boat.

If whaling work welded together men of the ship, whaling life linked together men of the trade. Isolated on the ocean's expanse, whaling ships sought each other out for company, and visits at sea, called *gams*, created a deep-sea society. A shipmaster who wanted to gam signaled his wishes to another captain by hauling his main yard back. If both agreed to a meeting, they hove to and exchanged boats' crews and captains. "The time is jenerally spent," claimed William Allen, who had just finished a late-night, Pacific Ocean gam with two other whaleships, "by the Capts in spinning the greatest lies about their personal engagements with whales – how far they can throw a lance and kill a whale. In the forecastle in singing songs none of the most moral and speaking about those of their acquaintants who had been killed or maimed by whales . . . descriptions of the different islands and ports they had visited &c of women & wine whoring & hard drinking."[67]

Another sailor likened the gam to women's get-togethers, so replete was it, he said, with gossiping and chatter: "Talk about a 'sewing society,' or a 'tea party' why it cant begin [to compare] with a whaleman's 'gam.' Everyone gets up their 'fantackle' and then they go at it, and the way small talk suffers is a caution. . . . And so the many tongues wag merrily for the space of three, or four hours when the command of 'man the boat' puts an end to the visit."[68]

Gams brought more than sociability to weary seamen. Gamming sailors cemented ties (if only temporary ones) by exchanging small gifts, such as tobacco, pipes, magazines, and books. Officers shared stories of successes and failures, exchanged information about whaling grounds, and gossiped about family and friends.

Gams were not perfect forums for fraternity among these men. Some sailors preferred to sleep than to socialize; others would rather have been looking for whales. And gams honored rank aboard ship. But these social meetings did unify the men of the fishery and at the least gave them a common source of pleasure. William Abbe, the sailor who felt so dissociated from his shipmates and who longed for the company of "true sailors" in merchant ships, also acknowledged the community that gamming engendered. As he was cruising on the offshore grounds in 1859, his ship took part in a triple gam, and in describing the scene of that gathering, Abbe waxed poetic. The three gamming whaleships, he wrote, "lay nearly in a triangle – on a quiet sea – in nearly a dead calm – while the moon at the full – shed a brilliant light over the ocean – & shone with elfish gleam on the white sails of the three ships – We could see in the distance

[67] William Allen diary, 15 September 1842.   [68] Ives diary, 28 January 1870.

two other ships gamming – & amidst this scene of quiet beauty – beneath the moon – joining with the new boats crew we danced away, sanding the decks & kicking off our shoes as we formed two cotillion parties & kept the decks alive – crowded as they were – with our shouts & laughter & music – I was lady to a stout negro – who laughed till he was hoarse & we all laughed & sang. . . ."[69]

American whalemen found common ground in the sociability of a gam, and they shared the ardors of work. They were also global tourists together. As they sailed to the earth's far horizons and scoured foreign oceans for their prey, they saw sights that distinguished them from stay-at-homes ashore and united them in awe. Whalemen exulted in their special perspectives on the natural world – in their unique views of such phenomena as waterspouts, lunar rainbows, the Southern Cross, the Magellan clouds, and, periodically, comets.

"Anyone that goes to sea can see a great many things that those who do not go never see," boasted James Knapp. "Now a sunrise to home is nothing compared to a sunrise at sea." Even the sight of that animal that whalemen regularly caught, cut up, and tried out could be an object of incomparable beauty. Samuel Chase challenged the "whole continent of America to produce a scene as strikingly sublime as a pod of whales spouting on a glassy calm sea. . . . Indeed, the sea appeared like a forest of ceders, such a multitude of spouts were. . . . heaving up into the air."[70]

American whalemen of all sorts celebrated their worldly authority by dismissing landsmen in general and by deriding land-based "authorities" of the sea in particular. Sailors who had brought with them ideas about pelagic beauty shaped by Byron or Coleridge or Wordsworth discovered the true nature of the ocean once they saw it close at hand. Along with their shipmates, they decried the pretty fantasies of men who had never really sailed on deep water.

Sarcasm is found repeatedly in the diary entries of the disillusioned. George Blanchard learned the hard way, for instance, that sailing in a gale could be less than awe inspiring. When his vessel encountered a violent storm in the Atlantic in 1842, the ship's deck was frequently a foot deep under water. Everything, he said, was "topsy turvy." Two casks of coal wrecked one of the bulkheads and "came tumbling into the steerage smashing the Berths and chests." The night was "as dark as a stack of black cats and blowing like the Devil," and the Divinity, it seemed, had nothing to do with the scene. This, Blanchard said, is "what I call fun alive." The following day, when it "blowed a screamer," and a heavy sea washed over the ship and "capsized" the crew's breakfast, he paid homage to the poets. "Oh!" he exclaimed, "the beauties of a Gale of wind at Sea."[71]

[69] Abbe diary, 17 June 1859.
[70] Diary of James Knapp, *Merrimac*, 18 March 1854, MSM; Diary of Samuel Chase, *Arab*, 19 December 1842, MeHS.   [71] Blanchard diary, *Pantheon*, 12–13 December 1842.

Sailors found figurative language particularly ripe for debunking. Those who expected the ocean to fulfill its literary promise as "mountains high" were annoyed when they did not find actual peaks of water. Orson Shattuck, greenhand aboard the *Frances* in 1851, wanted gullible readers to be warned about this misrepresentation. "I will state here that when people on shore read or hear about the waves being like mountains they may put it down as figurative language for they never are except in shape they are never more than 25 or 30 feet in height." George Blanchard likewise revealed that the highest ocean swells he had ever seen were 50 feet tall. "Oh these authers!" he exclaimed. "Their stories of old Neptune's Home are 'all in my eye.' "[72] In his dismissal of literary experts, Blanchard joined other sailors who debunked domestic projections and proclaimed themselves to be inside authorities on deepwater "reality."

This castigation of landsmen for their Romantic or naive portrayals of pelagic experience was often tinged with other bitterness. Victorian seamen, for example, who first came to sea exulting in gales and storms, became more cynical after working as forecastle hands. Repeated exposure to cold wind and salt spray and exhausting stints on the lookout or at the wheel quite miraculously transformed celebrations of God's brutal power into a denunciation of the captain's. Robert Weir, for one, came to a fast disenchantment on August 24, 1855. When he got "wet and tired out tending the rigging and sails," he tumbled into his bunk "with exhausted body & blistered hands." His final comment that day was an ironic one. "Romantic," he wrote.[73]

The circumstances that worked on Robert Weir to diminish the glamor of the sublime affected other men before the mast. In fact, although whaling work, social isolation, and oceanic travel brought forth a community of all hands, foremast men were driven to special consensus and alliance by a dictatorial shipmaster given to what sailors termed "hard usage." Collaboration might be created by the fear of whales' flukes or forged by common experience, but it would be treatment by ship's officers – the afterguard – that would truly inspire unity and collective action.

Shipmasters, though, were not all autocrats nor uniformly mean-hearted. They were diverse individuals who, like sailors, carried with them to sea various imperatives. The degree to which they galvanized a ship's company either with or against them depended on how they responded to these imperatives and used their authority. To better understand these complicated individuals, their approaches to power, and, ultimately, the social dynamics of the ship as a whole, we turn our attentions temporarily aft in Chapter 4.

---

[72] Diary of Orson Shattuck (under pseudonym of Charles Perkins), *Frances*, 16 January 1851, Log 994, WMLOD; Blanchard diary, *Solomon Saltus*, 18 November 1845.

[73] Weir diary, 24 August 1855.

4

The "Old Man"

The Sea Captain's Split Personality

It was, said Captain Samuel Winegar, the "longest night that I Ever Experienced and the Most anxious one." Winegar did not exaggerate, for on the night before he had nearly wrecked his ship on the Arctic shore. September 9, 1859, had begun auspiciously enough for the Massachusetts sea captain. His ship, the *Julian*, took a large whale, and its carcass had been chained to the vessel in order to cut it in. But a strong onshore wind had arisen and pushed the ship, immobilized with its prize, closer and closer to the rocky coast. "We had now got so neer," reported Winegar, that "we could see the huge Breakers through the Darkness of the night a Smashing up a gainst the rocks and a roaring like thunder threatening Sertain distruction to Everything that got in its way."[1]

Winegar knew that he might save the ship and its crew by unharnessing the whale and setting sail. The mate knew so, too, and asked Winegar if he would not do something to avert certain disaster. The first officer put the matter bluntly. If the captain did not take quick action, "We should Surely go on shore . . . and Every Man [will] be lost." Winegar, however, stood firm: "I told him no and to Stand by his Anchor." Thanks to "the kind hand of providence," the *Julian* escaped disaster. The following day, reflecting on the near demise of his ship and his crew, Winegar explained his earlier intransigence. "I had [rather] go to the Devil with a whale," he asserted, "than go to New Bedford without one."[2]

Cut from some of the same human cloth as Captain Ahab and Captain Bligh, Samuel Winegar demonstrates the monomaniacal determination that is considered

[1] Diary of Samuel Winegar, *Julian*, 9 September 1859, YUL.   [2] Ibid.

85

Whaling Captain Edward S. Davoll, pictured here, began sailing when he was eighteen, and he commanded his first ship when he was twenty-five. That may have been the high point of his life, for by the time he was forty his whaling career had been marred not only by a shipwreck but also by charges that he had, in 1860, taken an American vessel to sea for the purpose of picking up a cargo of slaves in Africa. It was while he was under the threat of federal indictment that he died of typhoid fever in 1863.

typical of shipmasters.[3] Indeed, much historical literature readily connects the rank of sea captain with terms of negative absolute power: despotism, tyranny, autocracy.

Evidence from the whale fishery suggests that certain features of the captain's tyrannic profile do indeed stand true. Pressured by owners at home to achieve profitable voyages even at the expense of seamen and licensed by the state to correct sailor disobedience, shipmasters could be forceful taskmasters and harshly physical disciplinarians. Whaling captains, though, were not all corporal despots. Weighing heavily on these men were not only courts of law but landward institutions of humanitarian reform as well as the members of their home communities – especially female members. These parties cautioned shipmasters to use restraint in the use of physical discipline and urged them to govern men through persuasion, not coercion. Some captains sympathized, at least in part, with what they said.[4]

I

To achieve annual average profits of 45 percent,[5] New Bedford whaling merchants had to have been shrewd businessmen.[6] Certainly shipmasters, who were on the receiving end of merchants' exhortations, threats, and reprimands, were painfully aware that money was the bottom line in the whaling business.

New Bedford owners made it repeatedly clear to sea captains that the aim of whaling voyages was to procure "large and handsome" cargoes of oil and bone while avoiding "lavish" expenditures. These merchants used a variety of strategies to encourage masters to meet their expectations. They tried to cheer sea captains on to whaling success. Refusing for the most part to acknowledge that luck had much to do with whaling, they insisted that if a master took the right

[3] Or at least Bligh as we have come to know him. For a revised view of this man and his style of management, see Greg Dening, *Mr. Bligh's Bad Language: Passion, Power, and Theatre on H. M. Armed Vessel* Bounty (New York: Cambridge University Press, 1992).

[4] Valuable for its discussion of shipmaster control in particular and social domination in general has been Ronald Takaki, *Iron Cages: Race and Culture in 19th Century America* (New York: Oxford University Press, 1990), pp. 283–4. Takaki, drawing on Antonio Gramsci's concept of hegemony, describes ways in which Captain Ahab of Melville's *Moby Dick* establishes domination over his crew through "consent rather than coercion." See, too, T. J. Jackson Lears, *No Place of Grace: Antimodernism and the Transformation of American Culture 1880–1920* (New York: Pantheon, 1981), p. 10.

[5] On the average profit rates of New Bedford owners, see Chapter 1 and Lance E. Davis, Robert E. Gallman, and Teresa D. Hutchins, "The Structure of the Capital Stock in Economic Growth and Decline: The New Bedford Whaling Fleet in the Nineteenth Century," in Peter Kilby, ed., *Quantity and Quiddity* (Middletown, CT: Wesleyan University Press, 1987), Table 10.22; and Davis et al., "Productivity in American Whaling: The New Bedford Fleet in the Nineteenth Century," *NBER Working Paper Series 2477* (1987): 12.

[6] See Chapter 2, footnote 4, for a list of the New Bedford owners studied in this chapter.

attitude and had sufficient ambition, skill, and hard-nosed savvy, he would
be home in no time with a full ship. "We think," wrote Charles R. Tucker to
Captain Charles Starbuck in 1836, "you will obtain [oil] with due energy and
perseverence . . . you have a good set of officers and crew . . . and you will find
them an easy set to manage." The confidence that owners exuded from the
counting house was such that on one occasion agents told a shipmaster who
complained about a leak in his vessel that, from their perspective, the leak "will
not be likely to increase." It would be good, they suggested, if the captain at sea
took the same positive attitude.[7]

Owners also encouraged competition among shipmasters and did their best
to stir up rivalry. "Captain Crapo," Swift & Allen wrote to one of their masters
in 1858, will soon be in the Pacific "to compete with you." To Captain Benjamin
Boodry, who reported that his vessel had taken 800 barrels of oil in 1857, the
owners wrote back that most of the other ships in the Okhotsk Sea that year had
taken 1,000. "[We] hope you had a chance to bring it up to [the] average of the
fleet before you left," they commented. These owners dismissed complaints
from masters with the same approach. When Captain Boodry balked at having
to recruit at Talcahuano instead of in Hawaii, the agents argued that Boodry
was out of line with his fellow captains: "All but you have liked it [Talcahuano]
much better [than Hawaii]." They evidently thought a personal comparison
might work to their advantage as well: "Capt Green got all he wanted at very
reasonable prices only about two months later than you were there."[8]

Merchants tried other means to bring out the best efforts of their captains.
They practiced the fine art of shaming, for instance. Swift & Allen had consistently pleasant communications with some of their sea captains, complimenting
them on success or congratulating them on their fortitude in the face of hard
luck, but they also knew how to berate a man. "The ordeal through which we
have passed in consiquence of your lavish expenditure last voyage," wrote the
owners to William Earl in 1857, "has certainly been a very unpleasant one to us.
Should the prediction of one of your owners be fulfilled (viz that you will do the
same again) it would be a lasting disgrace to you, and a mortification to us."
Asking Earl to draw on "all that is manly or honorable" within him, the owners
demanded his "utmost exertion" to redeem his reputation.[9]

Agents encouraged captains, prodded them, and humiliated them. They also
appealed to these men through perhaps their most vulnerable spot: their

[7] Charles R. Tucker to Captain Charles Starbuck, 20 June 1836, NBFPL; Swift & Allen
to William Hathaway, 5 August 1857, in "Swift & Allen Collection," Mss 5, S-g 3, S-F,
S-s 2, vols. 1-4 (letterbooks), WMLOD.

[8] Swift & Allen to James E. Stanton, 17 July 1858; Swift & Allen to Benjamin Boodry, 19
December 1857; Swift & Allen to Benjamin Boodry, 30 July 1857; For an example of
another shipowner who used the same tactics, see the letterbook of Matthew Howland of
New Bedford, 1858-1878, BLHU.

[9] Swift & Allen to William Earl, 13 November 1857.

pocketbooks. "You understand," wrote Jonathan Bourne, Jr., to Captain Isaiah Purrington, that "unless you obtain them thousand barrells in two seasons you do not get your bonus which is a further inducement to stay the third season." A year earlier, Bourne had cut to the quick with Captain Martin Palmer. Urging him to remain at sea for an additional year, Bourne noted that "you will add significantly to your property to enable you to remain at home with independence of feeling."[10]

Owners were even more hard-hitting during times of economic depression. The country's volatile financial climate at midcentury and during the Civil War, when the supply of officers frequently exceeded positions available, gave agents special opportunities to remind new or unproductive masters of the insecurity of their jobs. The partnership of Jireh Swift and Frederick Allen was particularly adept at informing shipmasters of the tenuousness of their employment. "The financial affairs of this country are in a tremendous vortex at the present moment," the agents wrote to Captain William Earl during the depression of 1857. "We shall loose some 15,000 on oil sold . . . which we hope will add another inducement if it is needed for you to bring us both a large and handsome cargo of oil." All the men, they added later, "who have not come up to the mark either in whaling or expenditures" have been "put back a step." They reminded Captain James Stanton, likewise, that "captains and officers are very plenty many that would have got a chance in former years will have to lay over, while those who go, get longer lays." Perhaps Captain Stanton did not or was not able to heed such admonitions, because five years later, during the Civil War, he was used as an object lesson himself. "Captain Stanton has opened a country store some two miles from Union Street," the agency warned a shipmaster, "but we think it would be closed at once if he had a good ship offered him."[11]

What rested behind the threats of owners was economic power. Nearly half of the New Bedford masters sailing at midcentury were not owners of their vessels but employees. Those who were co-owners were often shareholders of no more than 1/16th. These master-owners were guaranteed some job security – they could not be removed without a "special reason" – but it was not unheard of for a majority of the owners to find that special reason and to remove a man from command.[12]

[10] Jonathan Bourne, Jr., to Captain Isaiah Purrington, 16 August 1856, "Jonathan Bourne, Jr., Business Records," Mss 18, S–G, S–s 2, vols. 1–3 (letterbooks), WMLOD; Bourne to Captain Martin Palmer, 18 July 1855.

[11] Swift & Allen to William Earl, 7 November 1857, 31 September 1858; Swift & Allen to Captain James E. Stanton, 17 July 1858; Swift & Allen to Captain Reuben Crapo, 9 November 1863.

[12] Appendix V describes the ownership status of nearly 300 New Bedford captains from 1851 to 1880. On the removal of owners, see the consular report of William Crosby, Talcahuano, Chile, 21 March 1859, Ship *Falcon*, "Knowles Family Business Records," Mss 55, S–g 2,

Given the unequal partnership of owners and captains, and owners' strategies of strong-arming, it is not surprising to find masters both desperately eager
to please their agents and despondent when they could not do so. If captains
had good news, whether it was that their crew proved strong, the ship sailed
well, the provisions were adequate, or that the whaling was good, they did not
hesitate to report it. "I have only time before the mail closes," wrote home a
pleased Captain William Earl, "to reach over sea and land to take a wag of your
flippers collectively and listen a moment (in imagination) to your congratula-
tions. . . . I am of course, 'Earl' of the Donell, 1,000 whale 50 sperm. . . . It is
a source of infinite gratification to me, and the same to you, no doubt, that
fortune has been so propitious."[13]

Other masters, less fortunate, wrote back to fault-finding owners with defen-
siveness or in anger. Captain James Stewart wrote to Knowles & Co. in 1858
that "I do not admire the style of your letter (of July 14th). We are the victims
of circumstances and are compeled to submit to the nature of the case." He
could "promise nothing," he said in another letter six months later, "and can
only do what *man can do*." Swift & Allen obviously received similar protests.
"We must confess," the company wrote to Captain James Stanton, "yours of
Sept 8 . . . somewhat surprised us; how a sensible man like you could distort
our language to such a degree as to extract censure from it, is beyond our
comprehension."[14] These owners were clearly skilled at following one rebuke
with another.

If some sea captains became angry or defensive in the face of bad fortune,
others grew depressed. The journal of Captain Clothier Peirce is a particularly
impressive litany of misery, and it reveals the extremes of despondency to
which a shipmaster might sink. Peirce sailed in the "Poor Old Bark Minnesota,"
a ship that seemed fated to failure. "Som luckey Ship Has Boiled the Last
One," complained Peirce in June 1868. "Never did I feel more Cast down this
is A dark Period in my life; why did I come whaling but for my own distruction;
I think my damnation is fixed now."[15]

Events in the subsequent months did little to convince Peirce that his fortune

S–F, S–s 3, Fol. 5, WMLOD. The policy with regard to removing masters was made law
in 1872. See Walter Macarthur, *The Seaman's Contract* (San Francisco: Jas. H. Barry, 1919),
p. 157. See also Kendrick Price Daggett, *Fifty Years of Fortitude: The Maritime Career of
Captain Jotham Blaisdell of Kennebunk, Maine, 1810–1860* (Mystic, CT: Mystic Seaport
Museum, 1988), p. 116. Daggett's history of this shipmaster provides important insights
into the power and responsibility of sea captains in the mid-1800s.

[13] William Earl to Swift & Allen, 22 November 1852 in "Swift & Allen Collection," Mss 5,
S–g 3, S–F, S–s 1, Fol. 6–76 (incoming), WMLOD.

[14] James Stewart to Knowles & Co., 21 October 1858, 3 April 1859, "Knowles Family Business
Records," Mss 55, S–g 2, S–K, S–s 1, Folder 1, *Mary and Susan*, WMLOD; Swift & Allen
to James E. Stanton, 16 December 1859.

[15] Diary of Clothier Peirce, *Minnesota*, 29 June 1868 to 28 October 1869, Log 98A, WMLOD.

had changed. On July 22, the captain asserted that the "Hand of Providence is against me." On August 18 he claimed that the "Lords Power is against the Peirce Family." On August 21, it was the "Hand of Fortune" that had turned on him. On October 23, 1869, after five months without a whale, Peirce declared that "NEVER DID I See the WATER Here LOOK SO DESOLATE As It Does Now. Not so much as A PORPOISE to be seen." And, to firmly convince the reader of his journal that he was indeed an unhappy man, five days later Peirce reiterated his feeling that "My Ruin is Certain."[16]

Captain Peirce was discouraged, but he was not as seriously depressed as a whaling master named Thomas Peabody. On June 3, 1854, Peabody asked his officers if "thay thought a man would be punished in the other world for makeing away with himself if he had nothing to hope for." Apparently he decided he would not be punished, or he decided he did not care, for after taking laudanum for two days, he shot himself in the face.[17]

Masters who did not give in to serious depression nevertheless might go to unusual lengths to please. Captain George Pomeroy, who was a spiritualist, sought professional assistance from another sphere. He communicated regularly with his "Spirit Friends and Gurdian Angles" who were "floteing around in [a] butiful home . . . and bascing in the warm brite sunlight." One sailor who had met him reported in 1860 that "the spirits tell him where to go to get whales." Either the spirits were right or Pomeroy was lucky, because he returned to New Bedford in 1861 among the top of the fleet in oil.[18]

Lacking supernatural assistance, other men pushed their ships and their crews as far as they could go. Captain Samuel Winegar, whose mishaps in the Arctic began this chapter, was not alone in carrying desperation to extremes. The captain of the *Roman*, working in the Arctic in 1853, also became obsessive about getting a whale. Despite uncertain weather and a close call when his ship was "intirly encircled with Ice," he was determined to stay where he was. "I want one more whale bad," he wrote in his journal, "yes I want a good many . . . we have got but one months salt meet in the Ship but if I could get whales I would not mind [staying] untill that is gon and thin live on faith untill we get in where we can get some more." Whether the *Roman*'s men had to survive on faith alone as they finished Arctic whaling in early September is not known, but it is clear that they were not the only sailors suffering. "There is at least 150 ships on this ground that have 3 whales and less," noted the same captain, "and all bin beetting about through thick and thin through Ice and Snow and rain

[16] Ibid.

[17] Diary of Beriah C. Manchester, *Morea*, 3–5 June 1854, Log 135, WMLOD.

[18] G. P. Pomeroy to his spirit friends, 22 April 1858, MSM; Diary of Lewis Williams, *Othello*, 16 May 1860, PMS. Figures on the Pomeroy voyage are found in Alexander Starbuck, *History of the American Whale Fishery* (1877; reprint: Secaucus, NJ: Castle Books, 1989), p. 551.

among the rocks and off shore." The worst of this situation, he felt, was that despite their difficulties, captains like himself would "be blamed for not getting more Oil. . . . I think I hear some smart Ship Owner say, 'Tush I could get more Oil in Middleborough Ponds.'"[19]

This shipmaster worried more about facing his employer after a bad whaling season than he did about using and abusing sailors in pursuit of his prey. A glance at the letters owners sent out to captains may explain why this man ordered his priorities the way that he did. Whaling merchants, who saw labor costs as one of the biggest obstacles to money making, made it absolutely clear to captains that men before the mast could be seen as expendable. Aware that the labor supply at home and in many ports gave them some license in this regard, they became hardheaded when it came to sailors' interests and well-being.[20] As long as a seaman was in debt to the ship, through the outfit he had been given or through some other advance, owners wanted him to stay with the ship, by physical force if necessary. But as soon as the ship took whales and that man was due some money, owners asked captains to "encourage" him to leave. "Get clear of men whose room is worth more than his company," might well have been an agency slogan, so often did Swift & Allen urge it on their masters. Writing to Captain Reuben Crapo in the spring of 1858, these merchants were direct in asking him to "get clear" of any man who drained the company's profits "in the easiest way possible" or to "make him run away." They left the captain to decide how to provoke a sailor to desert. Agents similarly recommended (albeit with more subtlety) the pruning of crews toward the end of voyages when ships might be full and when sailors were due money. "Do not bring home any more men than is necessary to lower two boats," warned Swift & Allen to one captain. Charles R. Tucker concurred: "We should advise your coming home with only enough to manage the ship comfortably, thinking it a needless expense to ship a full complement to return home as many ships do."[21]

On occasion humanitarian concern did filter into owners' directives. One agent urged his masters to "promote peace and harmony," to "take good care of [sailors'] health," and "to give them enough to eat." A second (Quaker) owner insisted that captains should give "good treatment to the crew" and that they should "forbear knocking [sailors] about." Furthermore, he urged captains not to swear at sailors or call them "by other than their own names," or treat them "in any other way than becomes men." They should live, he said, "like broth-ers."[22] Paternal analogies were far more pervasive than fraternal ones, however,

[19] Diary of master, *Roman*, 22 May 1853; 21 August 1853; 13 September 1853, NBFPL.
[20] Labor markets in foreign whaling ports are discussed in Chapter 6.
[21] Swift & Allen to Captain Crapo, 6 May 1858; Swift & Allen to Captain Cornell, 3 August 1857; Charles R. Tucker to Charles Starbuck, 20 June 1836.
[22] Charles R. Tucker to Charles Starbuck, 20 June 1836. George Howland, Sr., to Captain Joseph Bailey, 5 May 1836, "The Howland Collection," Mss 7, S–g 1, S–D, vols. 3–4

as other owners directed sea captains to physically correct the young men under their care and to discipline them severely. One merchant urged a shipmaster to enforce the "strictest discipline," to make men "love and fear" you, and another advised a captain not to hesitate in reprimanding men swiftly: "Put them down at once," urged Charles W. Morgan, "Do not spare them."[23]

With some exceptions, then, what shipmasters heard from whaling agents were variations on the theme of savings and income, and, with regard to seamen, suggestions for stern treatment. Reminders of a captain's own precarious position only accentuated owners' insistent demands.

Captains listened carefully to whaling merchants, but they had ears for other individuals and institutions as well. Looming large over a captain's conscience and his comportment, sometimes supporting owners' interests and sometimes the interests of sailors, were legislators and judges and the long arm of admiralty law.

Although historians of the admiralty dispute the ancient origins of sea law, many scholars agree that nineteenth-century British and American laws governing the rights and behavior of officers and seamen were derived in part from the Rules of Oleron, a code of medieval origins.[24] The Rules of Oleron and all the legislation influenced by it had, according to Walter Macarthur, one "chief characteristic" when it came to the laws of seamen. This was "the obligation which binds the seaman to his ship during the period of [his] voyage." British admiralty laws of the eighteenth century, the first U.S. act of Congress relating to seamen in 1790, and subsequent U.S. legislation – particularly the Maguire Act of 1895 and the White Act of 1898 – did not deviate from this legal course.[25]

Over time, however, there were some modifications in the way the state dealt with seamen who defied their obligations. The Laws of Oleron, for instance, were very severe against seamen who broke their voyage contract and deserted, specifying in one case that sailors "be marked in the face with a red hot iron, that they may be known, and be infamous for as long as they live."[26] The United States admiralty system was kinder to deserters' faces, but throughout the nineteenth century it consistently supported the imprisonment of their bodies. In the notorious *Arago* case, which came before the Supreme Court in 1896, the court insisted that a sailor during the time of his employment had to "surrender . . . his personal liberty." The Thirteenth Amendment to the Constitution, which abolished "slavery and involuntary servitude," did not apply to

(letterbook), WMLOD; George Howland, Sr., to Captain John Munkley, 6 November 1843; George Howland, Sr., to Captain George Armington, October 1847.

[23] Jonathan Bourne to Captain James Munroe, 8 December 1854; Charles W. Morgan to George C. Ray, 6 July 1833, NBFPL.

[24] Walter Macarthur, pp. x–xi; Elmo P. Hohman, *History of Merchant Seamen* (Hamden, CT: 1956), p. 4; Grant Gilmore and Charles L. Black, Jr., *The Law of Admiralty* (2nd ed.) (Mineola, NY: Foundation Press, 1975), p. 7.

[25] Macarthur, *Seaman's Contract*, p. xix.   [26] Ibid., p. xiv.

mariners who wished to leave their vessels.[27] Finally, in the early twentieth century, legislation passed that allowed deepwater merchant mariners to desert a ship without fear of being jailed.[28] But whalemen would continue to be penalized for jumping ship, for the abolition of imprisonment for desertion did not apply to men working under the lay system. Even when American commercial whaling died, this exemption did not, for it is still on the books.[29]

Nineteenth-century shipmasters were thus given legal license to pursue, punish, and imprison men for desertion. They were also allowed to discipline men who resisted or challenged their command. Their authority, according to the *Shipmaster's Assistant*, was "necessarily summary and often absolute."[30] By defining mutiny broadly, they could punish a man if he resisted "lawful orders" or if he refused or neglected "proper duty" or if he "assemble[d] with others in a tumultuous manner."[31] Furthermore, the law allowed masters to use physical discipline. Particularly in the early decades of the nineteenth century, captains had a legal arsenal of weapons to their advantage. Until 1835 they could flourish knives and pistols to underscore their commands, and until 1850 they could discipline by flogging. Even after the middle of the century, when flogging was abolished, they could employ a deadly weapon to "defend their authority" in certain circumstances. It was not until 1898 that corporal punishment was made fully illegal at sea.[32]

Shipmasters might use physical means to punish sailors for past impudence or mutinous conduct, or they might also use it as preventive discipline. According to sailor and lawyer Richard Henry Dana, physical "treatment" could be an effective and legal reminder of the master's supremacy. "It is not necessary," wrote Dana in *The Seaman's Friend*, "that the punishment . . . be inflicted to suppress the offence at the time of its commission. It may be inflicted for past offences, and to promote good discipline on board." Dana acknowledged that it would be helpful for the master to make the reference to "by-gone" acts "clear and distinct" or they "will be presumed to have been forgiven."[33]

---

[27] Joseph P. Goldberg, *The Maritime Story: A Study in Labor–Management Relations* (Cambridge, MA: Harvard University Press, 1958), p. 15; Bruce Nelson, *Workers on the Waterfront: Seamen, Longshoremen, and Unionism in the 1930s* (Urbana: University of Illinois Press, 1988), p. 12; Macarthur, *Seaman's Contract*, p. xx.

[28] Coastwise sailors, by contrast, had been protected from imprisonment for desertion by the Maguire Act of 1895.

[29] Gerald O. Williams, "Share Croppers at Sea: The Whaler's 'Lay,' and Events in the Arctic, 1905–1907," *Labor History* 29 (1988): 55.

[30] Joseph Blunt, *The Shipmaster's Assistant and Commercial Digest* (New York: E. & G. W. Blunt, 1837), p. 151.   [31] Macarthur, *Seaman's Contract*, pp. 132, 149.

[32] U.S. *v.* Lunt, Fed. Case No. 15643; Macarthur, *Seaman's Contract*, pp. 132, 148–9.

[33] Richard Henry Dana, *The Seaman's Friend* (Boston: Little, Brown, 1844), p. 192. The picture of a sea captain flogging a sailor for an incident long past in order to promote good discipline is reminiscent of practices in that other more coercive institution of the

At the same time that the United States government continued to license sea captains as corporal disciplinarians, it nevertheless showed a growing willingness to outlaw the most violent acts of shipboard correction. Responding to a broad-based public campaign to limit or abolish punitive violence, Congress enacted a series of laws that curtailed sea captains' disciplinary latitude. Legislation in 1835 made it illegal to "beat or wound" or inflict "cruel and unusual punishment" without justifiable cause, and the 1850 statute abolished the use of the legendary cat-o'-nine-tails.[34]

The government also intervened in matters beyond ship discipline. A Congressional Act of 1790 declared sailors entitled to "good and sufficient Provisions," including a certain amount of "flesh meat" and ships' bread. Vessels were also required to carry a chest of "fresh" medicine, and ships going distances, around the Capes for instance, were required to have aboard antiscorbutics such as lemon and vinegar.[35] In 1835, the government made it illegal for ships' officers to unjustifiably withhold "suitable food and nourishment." Those not complying were liable to be "punished by a fine of not more than one thousand dollars, or by imprisonment not more than five years, or by both." And in 1872 new legislation provided for food surveys and provision scales, and established recourses for complaints from seamen about provisions.[36]

The history of admiralty law in the nineteenth century points to a government increasingly concerned, at least nominally, with sailors' welfare, but it also indicates that the federal government considered some marine industries more deserving of protection than others. Whaling owners and masters were called on to comply with many of the new protective measures, but it seems that they were exempt from others of them.[37] The Shipping Commissioner's Act of 1872, for instance, which provided for a wide range of laws having to do with the treatment of sailors, excluded whalemen. And in 1898, legislation was passed to provide sailors with a suit of warm clothing and a "safe and warm room for the

nineteenth century: the plantation. Eugene Genovese describes how preventive beatings could be used to impress on enslaved people the importance of future obedience. See *Roll, Jordan, Roll* (New York: Vintage Books, 1976), p. 208–9.

[34] Macarthur, *Seaman's Contract*, pp. 132, 148–9; Myra Glenn, "The Naval Reform Campaign Against Flogging: A Case Study in Changing Attitudes Toward Corporal Punishment, 1830–1850," *American Quarterly* 35 (Fall 1983): 408–25; Harold Langley, *Social Reform in the United States Navy, 1789–1862* (Urbana: University of Illinois Press, 1967), pp. 131–206.

[35] I. R. Butts, *Laws of the Sea: Rights of Seamen, Coaster's & Fisherman's Guide and Master's & Mate's Manual* (Boston, 1854), pp. 13–14; Joseph Blunt, *The Shipmaster's Assistant, and Commercial Digest* (New York: E. & G. W. Blunt, 1837), pp. 164–5; Macarthur, *Seaman's Contract*, p. 157.   [36] Macarthur, *Seaman's Contract*, p. 69.

[37] The degree to which whalers fell under new admiralty laws is not always clear. I concur with historian Gerald Williams's impression that legislation pertaining to different seafaring industries and fisheries is "a tangled web." See Williams, "Share Croppers at Sea," p. 40.

use of seamen in cold weather." Again, this law "did not apply to fishing or whaling vessels." Limits on the amounts that captains and owners could charge for slop chest items were also put on the books in the mid-1880s. This was not, however, to be "construed to apply to vessels in the whaling or fishing business.[38]

In a final example, the federal government took over the support of mariners' hospitals in 1884, ending sailors' forced maintenance of such institutions. But whalemen continued to pay for most of their health care. According to Elmo Hohman, whaling crew accounts included charges for medical care and charges for medicines kept aboard ship. A medicine chest charge of $1.00 to $2.00 per head was standard.[39]

Although their legislative policies clearly weighed more heavily on merchant shipping than on whaling, American lawmakers sought to extend their jurisdiction into the politics and management of all American ships. They were not the only land-based group that sought to govern shipmasters and sailors. Emerging from the evangelical movement associated with the Second Great Awakening and affiliated with the rising business class were many religious activists and social reformers eager to establish their influence at sea. They hoped to save souls among the seemingly impious population of mariners and to instill in "loose and immoral" men new respect for self-government. They also hoped to bring sailors back under the safe auspices of a home. And, not least of all, they hoped to relieve sailors' physical suffering and to end coercive brutality on shipboard.

In the national effort to reform seafarers, the American Seaman's Friend Society (ASFS) was by far the most successful and far-reaching organization in this period. Founded in 1825 and supported largely by women and ministers, the ASFS took its most energetic stand in seaports in the United States and abroad. Seeking to convince sailors to be better Christians, it sponsored the building of churches, bethels, and reading rooms. Hoping to encourage sailors to become better Victorians, it built seamen's banks. The society also constructed boarding houses where sailors could stay, be fed and well clothed, and be clean, temperate, and quiet. It was best known, though, for its sailors' "homes" where men might once again feel cared for and feel the ameliorating influence of domesticity.[40]

The ASFS also pinned some of its hopes on the sailing ship itself. The organization produced Bibles and tracts to uplift mariners at sea and established

---

[38] Macarthur, *Seaman's Contract*, pp. 68–70, 188–9.

[39] Ibid., pp. 153–4; Blunt, *Shipmaster's Assistant*, pp. 164–5; Elmo P. Hohman, *The American Whaleman* (New York: Longmans, Green, 1928), pp. 260–4.

[40] Hugh H. Davis, "The American Seamen's Friend Society and the American Sailor, 1828–1838," *American Neptune* 39 (1979), 45–57; Glenn, "Naval Reform Campaign," pp. 408–25.

floating prayer sessions to elevate them when they were anchored in port. Knowing that the shipmaster was one of the best means of converting the sailor and that he needed attention as well, the ASFS targeted sea captains with special pleas. The society urged captains to substitute persuasion for brutal force. The practice of flogging, the ASFS claimed, was inappropriate for citizens of a republican country. It made shipmasters themselves despotic, and it discouraged sailors from developing their own self-mastery.[41] Sea captains should regard themselves, according to the ASFS, as "father[s] to their crew[s]" and should "feel responsible for [sailors'] temporal and spiritual welfare." "There will be," the society hoped, "no more cruelty among them, . . . they will only exact that which is just and equal. Their men will then look up to them with affection."[42]

In tandem with their efforts to discourage flogging and other corporal discipline, this society and others like it also sought the abolition of alcohol at sea. Contending that seamen who were drunk provoked captains into particularly violent actions, seamen's reformers made the dry ship and "temperance reformation" two of the central goals of their campaign.[43]

The ASFS directed its efforts at all seamen, whalemen included. There were similar organizations aimed specifically at the whaling industry, however. The New Bedford Port Society, founded in 1830 for "helpfulness to the sailorfolk," sought like other groups the spiritual and moral reclamation of seafarers, and championed the rights and virtues of the common whaleman. "The sailor is a man," asserted the Reverend Wheelock Craig of the Society in 1857, "and his soul, like ours, if rescued from ignorance and wickedness, will exist enraptured and glorified forever."[44]

Craig was a knowing fundraiser. In telling prospective donors, including whaling merchants, that sailors suffered and needed help, he diverted the blame from the shipowners in his audience to Dame Fortune. He asserted that it was sailors' "lot" that subjected them to "hard labor, hard fare, and rude accommodations . . . to bitter partings, frightful casualties, and friendless exile."[45] He was careful, therefore, to awaken humanitarian concern without stirring up the least bit of guilt.

---

[41] Glenn, "Naval Reform Campaign," pp. 413–15. Myra Glenn convincingly connects the seaman's reform campaign to other antebellum campaigns (such as the abolitionist crusade) that sought to reduce corporal punishment.

[42] *Sixth Annual Report of the American Seaman's Friend Society* (New York: George P. Scott, 1834), p. 26.

[43] Glenn, "Naval Reform Campaign," p. 414; *First Annual Report of the American Seaman's Friend Society* (New York: J. Seymour, 1829), p. 27.

[44] Wheelock Craig, "An Address on the Occasion of the Twenty-Seventh Anniversary of the New Bedford Port Society" (New Bedford: The Executive Board of the Port Society, 1857), p. 1.   [45] Ibid., p. 4.

Weighing on the consciences of shipmasters as much as the men who hired
them, the Congress that governed them, and the ministers who sought to save
their souls were neighbors, friends, and family from home. Shipmasters were
readily reminded of domestic social obligations not only by the mail they picked
up in port or on gams but also by the hometown folks they carried with them.
Up until the middle of the nineteenth century, whaling was still something of a
local industry. At least some of the men before the mast as well as many officers
were recruited from the New England coastal area, not far from a shipmaster's
own backyard.[46] Towns and islands such as Mattapoisett, Nantucket, Martha's
Vineyard, and New Bedford, which sent many men to sea, were small communities, and social relationships within them were close and familiar. Population
growth and economic change had begun to move many men and women in and
out of these locales, but they were still places where people knew one another
and kept an eye on one another.

Some of this social closeness extended onto shipboard and played an important
part in personal relations at sea. Local sailors discovered, for example, that they
had a familiar eye of authority on them and that reports of their behavior might
easily make their way back home. The shipmaster, too, of course, was subject
to domestic scrutiny. In gams and in port and after the voyage was over, sailors
with wagging tongues might give word of a master's misdoings, not only affecting his ability to recruit but also damaging his social reputation.

Captain Peter Cromwell was one whaling master who knew how important
it was to cater to the hometown crowd. Cromwell, captain of the *Reindeer*, received numerous words of warning from his wife, Susan. On January 13, 1854,
Susan Cromwell reported to her husband that she had received a social call
from a woman whose son was on the *Reindeer*. "She wanted," Susan reported,
"your special care over him; she told me if I or my children was sick to send for
her at anytime & she would render all assistance possible." "Try Peter," Susan
urged, "to be patient with your boys & if they do not succeed at first let them
try again." Seven months later a "Mrs McKensie" came to visit Susan Cromwell
on a similar mission. Her son apparently was "not verry healthy" but that "all
the favour she expected or askd [was] that you would attend a little to him if he
was sick." Susan hoped, once again, that her husband would "have patience."
Then in July 1855, a Mrs. Hatch stopped by. Her son, too, was with Peter
Cromwell. Susan hoped that since the Hatch boy had been with her husband on
an earlier voyage he would be "easy" with him and would "recruit [his] ship
that all may have plenty to eat." She hoped to "hear no trouble" after her
husband returned to shore. The message from Cromwell's community at home
was clear, but it may have contradicted the directives of the captain's employers. Cornelius Howland, one of Cromwell's owners, wrote to the captain that he

[46] See Appendix I and Appendix III.

was concerned about the ship's "small profits."[47] Susan Cromwell had reason to care about whaling profits as well, but she chose to interject into the ship's affairs a cautionary note: if Captain Cromwell did not attend to a humanitarian agenda as well as to business interests, he would pay for it within his small circle of acquaintances, not to mention within the intimate confines of his marriage.

II

Shipmasters left port, then, burdened with an often contradictory set of objectives, admonitions, and prayers. As they made their way farther and farther from American shores, they would weigh one interest against another, listen to the landward voices that still echoed in their heads, and make hard decisions as to whose hopes and which ideals they should set adrift and which they might accommodate. Then they would tailor them all to the circumstances of the moment and to the special demands of the ship's company.

Clearly affecting a master's bearing on shipboard was the concern for hometown opinion mentioned above. The voyage of Captain Nathaniel Jernegan of the *Eliza Mason* in 1854 is a case in point. Captain Jernegan sailed with at least five men from his home on Martha's Vineyard. Jernegan knew these men or at least knew of them. They were, perhaps, part of that echelon of whalemen who were expected to stay in the industry and rise to its command.[48] They had certainly come with good recommendations.

Captain Jernegan soon discovered, however, the disadvantages of sailing with local men. John Pease, the second mate and a Vineyard man, proved to be a serious trial. He regularly fell asleep in his watch on deck, he was unable to manage the crew, and he developed an obsessive interest in Mrs. Jernegan, who had sailed with her husband. Every time she came on deck, reported a boatsteerer, Pease left what he was doing and followed her around to talk to her. "It matters not what he is employed [in]," the diarist explained, "he will go aft and engage in conversation until she goes below." (Although the captain's wife did not know it, the officer's infatuation took a more dangerous turn in his time off

---

[47] Susan Cromwell to Peter Cromwell, 13 January 1854; 29 August 1854; 15 July 1855, DCHS; Cornelius Howland to Peter Cromwell, 17 January 1855, DCHS. Lisa Norling discusses how whaling agents extended paternalistic assistance to officers' families but how such assistance "was shaped and limited by market imperatives." See "Contrary Dependencies: Whaling Agents and Whalemen's Families, 1830–1870," *Log of Mystic Seaport* 42 (Spring 1990): 9.

[48] Both Daniel Vickers and Lisa Norling describe the presence of upwardly mobile local men in the officer ranks of the whaling industry. See Vickers, "Maritime Labor in Colonial Massachusetts: A Case Study of the Essex County Cod Fishery and the Whaling Industry of Nantucket, 1630–1775" (Ph.D. diss., Princeton University, 1981), pp. 284–93; Norling, "Contrary Dependencies," p. 4.

A group portrait of unidentified New Bedford women, taken just after the Civil War.

duty. Apparently he "watched her in her room . . . to try and see her dressing and has made his boasts that he has seen her person exposed." His curiosity even extended to her menstruation, for he boasted of "seeing her underclothes stained and that she had her monthly turns.") Captain Jernegan eventually dismissed the officer but was reluctant to punish him. The boatsteerer thought he knew the reason. The "Capt endure it all," he explained, "because they [are] afraid his ma would feel bad or that some of the old grannies, and young misses in pantaletts in Edgartown would say that Capt J was too strict and too severe and did not treat him well."[49]

Captains who worried about sailor reporting occasionally tried to obstruct seamen's correspondence. Captain Richard Weeden of the *Ceres* attempted to prevent sailors' letters from going home. A foremast hand explained that while the *Ceres* was in Honolulu the captain refused to take mail ashore "for fear we had been writing about him." Weeden worried about hometown repercussions

[49] Diary of Orson F. Shattuck (under pseudonym of Charles Perkins), *Eliza Mason*, 20 December 1854, 18 December 1855, Log 995, WMLOD.

later in the voyage as well. After an altercation with a passenger who asserted that the Captain's behavior "would reflect no credit on him at home or any other place for such abuse," Weeden was apparently "nettled very much." And it seems he had further reasons to fear coming home. When he flogged a seaman on July 15, 1836, a shipmate who witnessed the beating commented that the captain "will hear of this inhuman treatment when we get home."[50]

Richard Weeden sought to stop the flow of unflattering reports by interrupting sailors' mail. Other masters took more preventive measures to silence sailors. They shipped men who knew nothing about the masters' hometowns and little of the people who lived in them. Most of these whaling captains, in fact, recruited foreign sailors at some point in their global rounds and could thus take cover in their anonymity. Whaling masters sailing in the later nineteenth century, when more and more of a departing crew was foreign-born, were especially safe from home-based scrutiny.[51]

For some masters, being safe from the prying eyes of hometown citizens, male or female, was not an issue. These were captains who devoted their ships to domestic reform and who might have welcomed aboard a band of female inspectors. These were captains who cared deeply about seafaring's potential for "barbarism" and who tried to turn their ships into polished models of Victorian discipline, restraint, and even religious devotion.[52]

It is difficult to estimate the number of shipmasters who carried with them full-blown commitments to self-discipline and sobriety, but it was not a tiny figure. The American Seaman's Friend Society described whalers in 1845 as among the most reformed of the nation's sailing vessels. They claimed as evidence the fact that "in the fisheries (including the whaling business) almost no liquor is used."[53] Although such assertions were hopeful exaggerations, it seems clear that temperance aboard whalers was on the rise. In contrast to the days in which sailors received daily rations of grog and when "whiskey and rum

[50] Diary of John Cleland, *Ceres*, 19 April 1835; 6, 15 July 1836, MM.
[51] See Appendix III and Hohman, *American Whaleman*, pp. 51–52. Merchant seamen and masters shared the opinion that localism was both useful and disadvantageous. Henry Davis, a merchant sailor in the 1850s, asserted that he would never again sail with a man that he knew, so that his behavior would not be "brought up when we get home." But he also noted that his shipmaster, who was "affraid we are talking about him," faced similar difficulties and that "those that live in glass houses must not throw stones." Davis said that his captain solved *his* side of the dilemma by shipping "a crew of South Sea islanders who cant talk a word of English." See Diary of Henry Davis, anonymous merchant vessel, 20, 21 August; 8 September, 1850s, MSM.
[52] Glenn, "The Naval Reform Campaign," p. 412, quoting *The Sailor's Magazine*, 1832, and an address to the American Seaman's Friend Society, 1888.
[53] *Seventeenth Annual Report of the American Seaman's Friend Society* (New York: Joseph H. Jennings, 1845), p. 13.

constituted a regular . . . part of every outfit," the antebellum period was relatively "dry" in the whaling industry. After 1840, according to Elmo Hohman, the number of ships sailing without intoxicants "became a clear majority."[54]

Both whaling captains and sailors themselves gave happy testimony to the dryness of some of the whaling fleet. Richard Hixson was pleased to be on the *Maria* in 1834 where, he said, all was "peace and pleasentness" because "ardent spirits" were forbidden. Sailors aboard the *Midas* "manifested a desire to join" in a "Total Abstinence Society" in 1842. All but three men signed the pledge aboard the *Tuscaloosa* in 1844 and threw overboard their "privet stock." Finally, on the *Eliza Adams*, seamen toasted sweethearts and wives not with grog but with a "filter full of cold water." This practice, remarked a less-than-sympathetic seaman, made a "fellow shiver and turn in as soon as possible, forgeting wives and everything els."[55]

Shipmasters who were not teetotalers were nevertheless often careful in the dispensation of alcohol. They frequently rationed their rum, offering a cup as a reward for taking and trying out whales or for surviving the rigors of the Capes, and they sometimes sent forward a tumbler of spirits in honor of the Fourth of July or New Year's Day. At other times they kept the liquor locker closed.[56]

Shipmasters not only instituted temperance at sea; they also encouraged or mandated piety. The captain who held abstinence pledges in one hand often carried Bibles in the other, and he was eager to share his possessions. The *Julian's* master emerged on deck in June 1847 with an "armfull of Bibles and Testaments to distribute among the men." The captain of the *Brunette* passed out "religious Books" to read that "were readily excepted of." Some masters did more than send reading matter forward. They assembled all hands for worship on the quarterdeck, and they conducted Bible classes. A few even formed Sunday schools. The cooper of the *Esquimaux*, the site of one of the Sabbath schools, noted that six days after the school was formed the crew held a religious meeting, and the "Lord was with us in an unusual manner." One sailor, touched by either God's spirit or the crew's, was overcome with emotion, and with

[54] Hohman, *American Whaleman*, p. 136. Many of the whaleships in this sample were not temperance ships. Of sixty-four diarists who give some indication of alcohol consumption on their ships, forty-nine (or 76%) described drinking on their ships, and fifteen (or 24%) described some efforts at temperance (generated by crew or afterguard) while the ship was at sea. But the presence of alcohol on some of these vessels does not diminish the argument that temperance was on the rise in the industry as a whole.

[55] Diary of Richard Hixson, *Maria*, 21 October 1834, HLHU; Diary of Avery F. Parker, *Midas*, 6 August 1842, Log 584, WMLOD; A. G. Goodwin to Swift & Allen, *Tuscaloosa*, 20 December 1844; Diary of John Jones, *Eliza Adams*, 3 January 1842, KWM.

[56] Diary of Eliza Nye, *Sylph*, 28 July 1847, MeHS; Diary of Albert Buel, *Java*, 1 January 1858, PPL.

"tears of repentance . . . down his cheeks requested an interest in our prayers that the Lord would convert his soul."[57]

The American Seaman's Friend Society kept an ear out for tales like this, for such stories gave credence to their claim that there was hope for sailors after all. They noted in their annual reports that an increasing number of shipmasters honored the Sabbath and that more and more whaleships, "however strong the temptations," refrained from taking whales on the Holy Day. In 1845 a particularly compelling testimonial came to their attention. One Sunday, it seems, sailors aboard an American ship spotted several whales cavorting close at hand. But the whalemen had just passed a resolution to honor the Sabbath. "We knew," said one mariner, "that we had taken very little sperm oil and that these monsters within our reach were worth from $2,000 to $3,000 each; that we, at least some of us, had families to provide for, and that a storm might come up in the night and take the bread from our children's mouths." Fortunately, these men did not have to make hard decisions between hungry children and God's grace. According to the society, they not only honored the Sabbath but returned to the United States "with a full cargo."[58]

Sailors themselves were occasionally eager and willing participants in programs for self-improvement and salvation. There were some men before the mast who experienced joyful conversion to the cause of Christ and who spent their dogwatch time earnestly contemplating their sins. But there were many men for whom the Bible and prayer books were not reading matter of choice. Ezra Goodnough, aboard the *Ann Parry* in 1847, joined his shipmates in reading the Bible because "we have read evrything else there is in the ship so much that we have it all by heart and we have talked over everything that has happened for the last ten years so we are read out and talked out." Resorting to the Bible was a clear sign, Goodnough said, that "it is time to go home." Like Bible reading, prayer meetings and Sabbatarianism may have recruited sailors with mixed motives. Sailors favored honoring the Sabbath because Sundays meant "Better fare & less work," or a day when they might "break out their chests and [air] their clothes and [view] their bills which is generally enourmous." Alternatively, they might use the time to find a quiet place to pursue "some vile novel." The contemplation of sin and salvation may have played a small part in the activities of this Holy Day.[59]

---

[57] Diary of F. Cady, *Julian*, 27 June 1847, KWM; Anonymous diary, *Brunette*, 24 July 1842, PPL; Diary of J. Hersey, *Esquimaux*, 18 March 1843, KWM. See also Diary of J. Hersey, *Shylock*, 22 June 1851, KWM; Cleland diary, 8 May 1836; Hixson diary, 21 October 1834.

[58] *Seventeenth Annual Report*, ASFS, pp. 13–14.

[59] Diary of Ezra Goodnough, *Ann Parry*, 22 August 1847, PMS; Diary of Isaac Jessup, *Sheffield*, 2 September 1849, MSM; Diary of Robert Weir, *Clara Bell*, 23 September 1855, MSM; Diary of Silas Fitch, *Charles Phelps*, 4 February 1844, MSM; Diary of Charles Stedman, *Mt. Wollaston*, 12 February 1854, NBFPL.

The practice of temperance and piety at sea was sometimes inseparable from conflicts over power, discipline, and living conditions.[60] Shipmasters rewarded pious and obedient seamen with special favors, and sailors "put on" piety hoping for better treatment. Men aboard the *Charles Phelps* "testified to the truths of the gospel" in return for liberty on shore. Whalemen aboard the *Ceres* attended prayer meetings when their meals were adequate. Seamen of the *Eliza Adams*, hoping to please the captain, "lay their bible or testament by the side of them when they go to sleep to make the old man think they had ben reading it." Such ploys were not certain to fool the shipmaster. When Captain A. Goodwin wrote to his owners that his men had signed the pledge, he assumed that the men had an "eye fixt on a higher berth," not that they had had a real conversion to the cause of temperance. Either way, he got the behavior he sought.[61]

If some shipmasters negotiated with sailors over social behavior, others employed more despotic approaches to morality. They threw playing cards overboard, they forbade swearing and loud noise, and they punished men for dancing on the Sabbath. Sailors, for their part, were not always accommodating. Men aboard the bark *President*, for example, told the master, when he demanded their playing cards, that "they cost their money & thay considered the cards to be their lawfull property." The captain gave them until the next day to comply with his demands, but when junior mates armed themselves with scrub brooms and pikes to defend the foremast men, the captain backed down.[62]

Mariners in conflict with a "reformed" master could always look to the ultimate tool of protest – desertion – as a solution to their disagreements. Sailors aboard the *Sea Queen*, according to William Abbe, were all planning to leave that ship because of social differences. The *Sea Queen*, claimed Abbe, was deathly. Her decks were "silent & the forecastle dark." The ship had "prayers every morning . . . her skipper officiating as chaplain – No noise of singing or dancing or music is allowed – no loud conversation permitted." This ship's skipper was afraid to go into port, explained the whaleman, "for the very natural fear that his crew would desert him."[63]

There were evidently many strict Victorians in the whaling aftercabin, but there were also some men who had little taste for temperance and piety. A few shipmasters drank (a lot) and allowed sailors to do the same. The *Atkins Adams* crew, for instance, made beer ("swankie") out of molasses, water, and potatoes,

---

[60] On power struggles ashore over the same issues, see Paul E. Johnson, *A Shopkeeper's Millennium: Society and Revivals in Rochester, New York, 1815–1837* (New York: Hill & Wang, 1978), pp. 138–41.

[61] Fitch diary, 24 April 1843; Cleland diary, 2–8 May 1856; Jones diary, 7 March 1852; A. Goodwin to Swift & Allen, 20 December 1844.

[62] Cady diary, 6 July 1847; Hersey diary, *Shylock*, 6 July 1851; anonymous diary, *President*, 11 January 1844, PPL.

[63] Diary of William A. Abbe, *Atkins Adams*, 17 June 1859, Log 485, WMLOD.

and the captain's only concern was that the whalemen were "using so much water." Some sea captains encouraged dancing and even delivered accordions or fiddles to foremast hands to facilitate noisemaking. And there were plenty of shipmasters who ignored Sundays "off soundings," particularly on whaling grounds. Captain Samuel Winegar made his views on this subject especially clear. In considering the piety of another master in the whaling fleet, he remarked that "Capt Kelly's Religeous scruples will not allow him to whale it on the Lords day he thinks it a Sin while I think it a Sin if a Man sees whales and does not try to ketch [them]."[64]

Some captains, then, chafed at temperance advocates and scoffed at Sabbatarians. Many also turned a cold shoulder to the voices that pleaded with them for forbearance toward their sailors. Captain Samuel Braley was one man who struggled to adhere to Christian injunctions against physical violence but who lost the battle. One Sunday at the end of November 1851, Braley knocked a man over "three or four times" because he had used "bad language." The captain, who had just completed a private prayer session in his cabin, was seriously disturbed by his own lack of restraint.[65]

But how was it, Braley asked himself, that a shipmaster could bring a crew to respect him if he could not punish them with physical means? How was a captain supposed to "requite evel with good"? He understood that corporal chastisement was "contrary to the precepts and example of the Holy Jesus," and, God knows, he had tried to be a Christian shipmaster. For the first two years of his voyage he had resisted punishing a single sailor. When men committed offenses, Braley had "called them, talked to and persuaided them to behave even with teers; and not make me punish them because I disliked to do it." And some men, Braley reported, had responded well to this approach. But others, it seems, began to suspect that he lacked manly courage. "It is not because he lackes the will," the captain imagined them saying, "but because he dares not do it."[66]

Faced with such a slight on his manhood and fearing a serious breach in ship discipline if he did not act, Braley rose to the occasion. He let his fists do the talking, and the result convinced him that Christian restraint probably had a place in the world but that place was probably not on a deepwater ship. "The laziest, most stubbornly uncooperative men in the ship," claimed Braley, were now the "most attentive and active men." "Such is the effect," exulted the captain, "of giving one man a pounding. . . . Every man is here to do my bidding, and to do it cheerfully too, and they shall do it!"[67]

Thus Captain Samuel Braley remained unconvinced when it came to applying

[64] Ibid., 26 July 1859; Winegar diary, 18 September 1859.
[65] Diary of Samuel Braley, *Arab*, 23–24 November 1851, KWM.
[66] Ibid.   [67] Ibid.

the tenets of Victorian Christianity to the administration of the sailing ship. This man was somewhat more attentive to admiralty laws, though, or at least to the men who were willing to test them and sue captains for damages. "There are always enough on shore to listen to jacks yarns," lamented Braley, "and put more reliance on it than they would in everything that the master and all his officers could say." Braley himself resisted punishing a sailor he suspected of malingering because he did not want to be accused of forcing a sick man to duty or, as the 1835 statute put it, of inflicting "cruel and unusual punishment." "Oh and alas," exclaimed Braley, ". . . if [he] says that he is not able to work I must not drive him to it; for although he mite be well and I have evry reason to believe that he was so at the time; he mite be taken realy sick in one hour afterward and then what a cry there would be raised against the Capt of the three masted ship So I must let it be for the present and wait. There has been many a master accused of killing men because he drove them to work when they were playing the "old Soger" and perhaps . . . the same man was taken realy sick and died. That was enough for Jack the old tyrant has driven that man into his grave."[68]

Samuel Braley was not the only master concerned about the litigative power of his seamen. Benjamin Neal, who worked in the Indian Ocean in 1839, confessed that he had to turn his ship over to his men in part because he feared legal reprisal. When he sailed into port, his seamen forced him to sleep ashore. Unsupervised, three of his sailors became "beastly drunk" and made "noise enought to have confused the worst Brothel in existance." But Neal felt that if he disciplined them, they would retaliate. "If I had struck them or used any harsh means," he protested, "it wold have bin a fine hande for them in Salem."[69]

A survey of legal cases resolved in Massachusetts federal courts from 1825 to 1861 gives us some indication of how much masters like Braley and Neal actually had to fear from sailor litigation.[70] Sailors who pursued lawsuits in court generally received reimbursements for court costs and won an award for damages. The amount of their awards, usually around 10 percent of the damages requested, averaged 50 to 75 percent of a master's monthly wage, not including court costs.[71]

---

[68] Ibid., 29 May 1852. The legislation quotation is taken from Macarthur, *Seaman's Contract*, p. 148.

[69] Diary of Benjamin Neal, *Reaper*, 8 March 1839, EI.

[70] See Appendix VI, Table VI.4. Whalemen, says historian Gaddis Smith, worked "almost exclusively" through the federal court system after the 1830s. Smith discusses the importance of legal records in studying whaling in "Whaling History and the Courts," *Log of Mystic Seaport* 30 (October 1978): 67–80.

[71] It is likely that many cases were settled out of court and that sailors recovered some damages informally. When a Captain Nichols was cleared of charges of assaulting a mariner in 1854, the judge, Peleg Sprague, was not simply being sympathetic to the master. It turned out that Nichols had paid off the sailor (to the tune of $50.00) and that Nichols had a receipt

The threat of lawsuits, then, may have served as a deterrent to some disciplinarians at sea. Black and white, literate and nonliterate Americans, as well as men with Portuguese surnames, took shipmasters and mates to court in Massachusetts.[72] What we do not know is how many aggrieved seamen did not pursue court justice. It was not necessarily a matter of money: lawyers often assisted impoverished sailors by taking cases on a contingency basis. But certainly some mariners could not afford the time or energy to seek formal damages or retribution. Others may not have survived the voyage and lacked descendants to sue on their behalf.[73] We also do not know whether men who hailed from countries outside the United States were as litigious as Americans and how accessible courts in foreign ports were to common seamen. Sailors' diaries suggest that shipmasters had less to fear from non-Americans and that captains exercised the greatest restraint when it came to men who could easily pursue a crime or an injustice in a United States court.[74]

Although evidence points to potential litigation as having at least some deterring effect on shipmasters, it also indicates that masters were not very fearful of criminal prosecution per se. Breaking one of the new laws against cruel punishment or against flogging did not seem to cut heavily into sea captains' peace of mind. Again, legal records may explain why. The monthly salary of New Bedford whaling captains averaged around $108 at midcentury. Fines handed to shipmasters convicted of cruel treatment at the same time were rarely more than half of this salary and commonly less than 20 percent of the monthly wage.[75] These amounts simply may not have been much of a disincentive.

Thus the justice system extended itself out to sea in uneven ways. Judges ashore were supportive enough of sailors' rights to explain the nervousness some masters felt with regard to ship discipline and judges' decisions should caution historians against uncritical depictions of shipmaster brutality. At the

from the sailor "exempting him of further charges." See Lobe v. Nichols, Special Massachusetts District Court, December Term 1854, National Archives, Waltham, MA. On wages, see Lance E. Davis, Robert E. Gallman, Teresa D. Hutchins, "Risk Sharing, Crew Quality, Labor Shares, and Wages in the Nineteenth Century American Whaling Industry," *NBER Working Paper Series* 13 (May 1990): 61. On awards for damages see Appendix VI, Table VI.4.

[72] Appendix VI, Table 6 describes the literacy of plaintiffs. Given the apparent poverty of some of the men suing for wages and damages in Massachusetts, it is likely that some lawyers waived fees until after cases were settled to their clients' advantage.

[73] Smith, "Whaling History and the Courts," p. 79.

[74] Nathaniel Saxton Morgan, diarist aboard the *Hannibal* in 1849 to 1850, claimed that his master beat a Hawaiian Island native with impunity because the man was "ignorant of the language and laws of the Country" and would have "no redress" (20 October 1849; 29 January 1850, MSM).

[75] For masters' salaries see Davis et al., "Risk Sharing," p. 61; for court fines see Appendix VI, Table VI.2.

A. A. Von Schmidt.   J. Halpin.

A PICTURE FOR PHILANTHROPIST.

Flogging, as a means of discipline and punishment on shipboard, was
abolished by the U.S. Congress in 1850. The campaign against flogging
and other corporal punishments was linked to the abolitionist movement
and its condemnation of the use of physical force in slavery. The anti-
flogging movement used illustrations like this one to help dramatize its
case.

same time, however, there were enough loopholes in admiralty law to allow
shipmasters considerable license.

In fact, diary evidence, along with additional court reports, reveals the many
occasions on which sea captains performed true to reputation. The nature of
physical discipline at sea was as wide ranging as a master's imagination and as
diverse as the weaponry at hand. In demanding and commanding obedience,
masters, and mates under their supervision, used various tools of the whaling
trade, or any "household" object. Belaying pins, bricks, clubs of firewood, and
even, on one occasion, a pudding bag, flew at sailors in heated moments.
Flogging and daggers and loaded pistols were outlawed by 1850, but this does
not mean that they were never used. Court cases in which a sailor sued a captain
over the illegal use of a whip or a gun are many and probably represent only a

fraction of the actual instances in which a dangerous weapon was actually employed.

Members of the afterguard took up firearms and loose objects to assert their authority; they also used the flat of their hands and the tip of their hard boots. By far the most common means of quarterdeck discipline was the punitive kick or punch. Shipmasters who grabbed sailors by the hair or beard and hit them in the face or groin swiftly reminded them of their inferior status. Less commonly, they approached sailors at the ship's wheel, where a defending mariner could do little to avoid a physical affront. Captain Gray of the *Hannibal* was one of several masters who chose to discipline sailors who were steering the ship. Calling a Hawaiian sailor at the helm a "damn whore's ghost," Gray kicked and struck the man repeatedly. When the sailor went off duty, his "body was all covered with ridges."[76]

Whaling masters and mates hoped that hard-hitting correction would render sailors submissive at once, but officers sometimes extracted subordination from seamen only after lengthy ordeals. Sea captains put recalcitrant seamen on week-long diets of bread and water, they imprisoned them in lockers and between decks, and they confined men in trypots. Sailors who hid away below decks could be smoked out, as men aboard the *Richmond* discovered in 1846 when the shipmaster decided to force them to work or send them all "to hell together." The men emerged and were variously shackled or flogged.[77]

If slow torture did not have the desired effect, additional physical punishment might be necessary. Captain Avery E. Parker of the whaleship *Midas* whipped a sailor repeatedly in 1843 until he eked out the proper words of submission. "I asked him again if he would answer me civilly for the future," explained Parker in the ship's log. "He replied he would answer me as he allways had. I told him he did not answer my question." So Captain Parker beat it out of him: "I gave him in all, four half Doz of lashes at intervals of ten or twelve minutes, asking him between each two half dozen, if he would Answer me civilly." Eventually the sailor's resistance was broken, and he terminated the cycle of whippings with a simple, "Yes sir."[78]

James Quick, seaman on board the *Roscius*, described a similarly brutal tug of war with the captain of his ship. In a suit for damages against the shipmaster, Quick asserted that he had been called aft from the forecastle and asked by the master, "Can you behave yourself," to which he had replied, "Yes Sir, I guess so." "You guess at it do you?" the captain retorted and allegedly drew back and struck Quick in the face and kicked him three to four times on the back side. He then made Quick sit on a spar for the better part of an hour and then put

[76] Morgan diary, 20 October 1849; 29 January 1850.
[77] Diary of Charles F. Morton, *Richmond*, 12 March 1846, NBFPL.
[78] Parker diary, 26 December 1843.

him below in irons. Later when the mariner complained that he was lame and could not do a full day's work, the master supposedly said he would "cure him" and hit him on the shoulders twelve times with a reef point and hung him in the rigging for two to three hours.[79]

Attempts to keep sailors in their place were sometimes accomplished with more subtlety than pure physical force. Shipmasters had a very effective means of stifling mariner resistance through the application of medical therapeutics. It worked this way: the captain controlled the ship's food supply, the length and course of the voyage, and the pace of work – all of which had direct bearing on whalemen's health. He was usually also the ship's doctor. The double role gave the master an incentive to promote the good health of seamen. It also, however, allowed him to dismiss complaints that resulted from his own shipboard policies and to "cure" a man any way he wanted to.

Candidates for a "cure" included men who seemed to be resisting work by pretending that they were unwell. These men were playing what was known as the "old soger." As a third mate explained it: "On board of a whaleship where we have No doctor the crew will try to take the advantage and play the Sick man." Their looks, he added, "most allways betrays them."[80]

Or so sea captains and officers thought, and they contrived a way to "fix" the miscreants. They were legally debarred from forcing a sick man to work, but, thanks to contemporary medical practice, they could make treatment synonymous with discipline. Many American physicians in the mid-1800s still practiced what is known as "heroic medicine." Heroic practitioners believed that illness stemmed from a systemic imbalance, and they attempted to restore health by aggressively adjusting body fluids and controlling intake and output. They bled, they purged (producing massive diarrhea), and they induced vomiting. They also applied blisters to the surface of the skin in order to produce a "healing" irritation. Through such intervention they hoped to bring the body back to what they believed was its natural state of equilibrium.[81]

Through such intervention a shipmaster or mate might also punish a malingerer. "The ownly way [to handle suspected malingerers]," argued officer Edwin Pulver, "is to give them the most disagreeable drugs which will make them sick

---

[79] Quick v. Winslow, Special U.S. District Court (Mass.) September Term, 1854. National Archives, Waltham, Massachusetts.

[80] Diary of Edwin Pulver, *Columbus*, 12 December 1851, PPL.

[81] By the 1830s heroic medicine was under attack in academic circles, with increased acceptance of the idea of self-limiting diseases and natural recovery. Yet, according to Charles Rosenberg, traditional heroic approaches had "extraordinary tenacity." See Charles E. Rosenberg, "The Therapeutic Revolution: Medicine, Meaning, and Social Change in 19th-Century America," in Judith Walzer Leavitt and Ronald L. Numbers, eds., *Sickness and Health in America* (Madison: University of Wisconsin Press, 1985), p. 46; and John Harley Warner, *The Therapeutic Perspective: Medical Practice, Knowledge, and Identity in America, 1820–1885* (Cambridge, MA: Harvard University Press, 1986), pp. 85–7.

any how wether they wish to be or Not." Pulver's captain was able to "cure"
two men who had complained of being sick. He gave one man a heavy dose of
"saults," and to the other, who had complained of a pain in his side, he
administered a blister of "spanish flies mixed with pepper sauce." When the
patient requested that the blister be removed or that he be treated with cam-
phor for the pain he suffered, the captain refused. Officers also tried emetics on
alleged malingerers. The mate of the *Lucy Ann* suspected on November 19,
1842, that a sailor named Ben was "shamming." A day earlier Ben had been
thrown out of a whaleboat and had suffered shock from the cold water. The
mate, though, believed that he had had sufficient time to recover and sent him
to work after giving him an emetic and making him vomit. He then "promised
to give him one every morning until he got well." On the *Gratitude*, a foremast
hand named Big Bill feigned illness "about twenty times . . . when any extra
hard work was to be done." Forced vomiting was again the treatment of choice.
"I reckon," said a boatsteerer who witnessed the therapy, "he wont soger much
more for he really is sick now."[82]

Heroic medicine was in many ways the shipmaster's perfect tool of correction. Neither the federal government nor humanitarian agencies could fault a
captain for violating seamen's rights when all he did was treat sailors with
accepted therapies. The insistent attentions of a shipmaster, no matter how
aggressive or painfully felt, could easily be explained as sympathetic concern.[83]

Afterdeck authority, then, might be achieved through physical means, either
"therapeutic" or more explicitly punitive. Shipmasters had additional strategies
for establishing command on shipboard. Knowledge is power, and whaling
masters sometimes withheld from sailors information about the ship's whereabouts. In contrast to captains of a century earlier, who worked with their crews
"within a broad system of apprenticeship," and who probably shared some
trade secrets with seamen, whaling captains had something of a lock on navigational information.[84] On these ships there were many second or third mates who
were not capable of taking the ship's position.[85]

[82] Pulver diary, 12 December 1851; Diary of John Martin, *Lucy Ann*, 19 November 1842,
KWM; Diary of Lewis Williams, *Gratitude*, 21 June 1859, PMS.
[83] After all, sea captains took emetics and cathartics themselves. One shipmaster even administered a cathartic to his constipated dog. See Diary of Edmund E. Jennings, *Mary and
Susan*, 24 August 1877, 2 November 1877, PPL.
[84] Marcus Rediker, *Between the Devil and the Deep Blue Sea: Merchant Seamen, Pirates, and the
Anglo-American Maritime World, 1700–1750* (Cambridge, UK: Cambridge University Press,
1987), p. 87.
[85] According to Richard Henry Dana (*Seaman's Friend*, p. 146), this practice was not unusual
on American deepwater vessels. Dana claimed that it was not even necessary for the second
mate to know navigation. See also Hohman, *American Whaleman*, p. 48. The widening
separation (and growing power inequality) between men at the top of the ship's hierarchy
and those at the bottom was also characteristic of Canadian merchant sail during the same

The sources examined here contain numerous references to masters who not only did little in the way of sharing skills but who also kept the ship's position a secret. Captain Weeden of the *Ceres*, who, according to a seaman, was "near going ashore" through his own faulty calculations, refused to let the mate "show sailors how to work lunars." The master of the *Cavalier* also guarded his whereabouts jealously. "The Old Man," remarked seaman William Wilson, "does not like that we should know the ship's position, and when the Steward looked upon the slate to see longitude, he asked 'what business he had to look at the slate?'"[86]

Men who sailed under commanders like these might never be sure where they were. Mariners aboard the schooner *S. R. Roper*, for instance, knew that they were around thirty miles from land on October 10, 1865, but whether this was the continent of Africa they were not sure because they were not "let into the secrets of the Capt & Mates Log-Book." Similarly, sailors on the *Atkins Adams* knew that they were bound into port but had virtually no idea to which particular place they were destined. "No one knows where we are bound, but reports of twenty different ports have been flying around the ship for the last two weeks," a sailor commented.[87]

Beyond mystifying their position and elevating their authority, shipmasters kept navigational secrets for at least two other reasons. On a practical level, their refusal to reveal whereabouts made it difficult for seamen to make detailed schemes for desertion and stifled sailors' early complaints about unpopular destinations. By not sharing information about their positions, masters also protected their own ignorance or incompetence. It is unlikely, for example, that the crew of the *Arab* knew what a quandary their captain was in when he was disabled by indecision. Samuel Braley, master of the *Arab*, was thoroughly stymied as to where to direct his voyage. "Away I go with a fine breeze," he remarked in January 1850, "but where am I bound; that is a hard question; one that cannot be answered. I have been racking my brain now for several days, trying to pitch on a place to go."[88] Thanks to the likely navigational ignorance

period. Eric Sager explains that in 1850 the British Parliament passed the Mercantile Marine Act, which, among other things, required the certification of masters and mates and the training of them in navigation. Parliament hoped, Sager claims, to further the safety of British vessels by enacting this legislation, but it also hoped to "elevate both captains and mates . . . ," to "arm them with proper power over their crews," and to "strengthen the reins of discipline." Eric Sager, *Seafaring Labour: The Merchant Marine of Atlantic Canada, 1820–1914* (Kingston, Canada: McGill–Queens University Press, 1989), p. 95. (Sager quotes from Hansard, *Parliamentary Debates*, 3rd. Series, CVIII, 11 February 1850, 668, 671.) The United States did not require the actual licensing of masters and mates until the turn of the twentieth century (Macarthur, *Seaman's Contract*, p. 15).

[86] Cleland diary, 25 September 1836, 18 October 1836, MM; Diary of William Wilson, *Cavalier*, 26 December 1849, MSM.
[87] Diary of Caleb Hunt, *S. R. Roper*, 10 October 1865, KWM; Abbe diary, 21 March 1859.
[88] Diary of Samuel Braley, *Arab*, 16 January 1850, KWM.

of his crew, as well as to the legitimacy of some meandering, Braley's dilemma was probably a well-kept secret.

Shipmasters who kept privileged information might direct successful, if highly autocratic, voyages, but such men might also lead a vessel and its crew to disaster. No fewer than a dozen captains in this sample drank to inebriation while at sea and became incapable of handling the vessel, much less setting its course. Some of these men found protective cover in the first mate, like the captain of the *Columbus* who took "freely of the water of life," and like Benjamin Clark, master of the *Chili*, who, according to his sailors, "got drunk often & when boozy would go aloft with two spyglasses & sing out to keep the ship steady."[89]

Less amusing and more lethal to his officers and sailors was William Heath, captain of the *Euphrates*. Heath, who drank steadily through the month of September 1860, had no man aboard ship who was proficient enough in navigation to take his place. It was all right, it seemed, as long as the man's liquor held out. But when his rum disappeared at the end of September, he went into serious withdrawal. Ambrose Bates, the mate and logkeeper, noted on September 28 that "Capt Heath is visited with the blue devils." Worst of all, Bates noted with alarm, was the fact that one of the "devils" seemed to be Bates himself and that the captain threatened "to shoot somebody."[90]

As the ship neared a "chain of dangerous islands" in the north Pacific, Heath continued to hallucinate. A crowd of "Rusian girls," he insisted, had boarded the vessel, and a coat on a chair in the aftercabin was a malevolent human. Although Bates himself was off the hook, he realized that the crisis was hardly over. "Something must be done," he declared, "for the safty of our lives and ship." After two tense days, the captain recovered somewhat, and he began to take control of the ship's course. Two weeks later, said the relieved first officer, Captain Heath was "enother man in apperiance," and "very interested in his business."[91]

By far the most frightening demonstration of how a lopsided system of knowledge and power might run amok for owners and sailors alike took place in the summer of 1871 in the Pacific Ocean. On July 27 the master of the *William Gifford*, Richard Veeder, interrupted a whaling cruise in the Marquesas Islands to stop on shore. There, according to a foremast hand, he performed a private ritual: he buried a child "in a flour bag." It was rumored that the child was his own, the issue of his liaison with a native woman.[92]

The child's burial was but one of a long string of strange occurrences that

---

[89] Pulver diary, 14–16 September, 1852, PPL; Abbe diary, 14 June 1859.

[90] Diary of Ambrose Bates, *Euphrates*, 28 September 1860, KWM.

[91] Bates diary, 29 September, 12 October 1860.

[92] Diary of Edward J. Kirwin, *William Gifford*, 1871–72, KWM. All subsequent references to Kirwin or to the *William Gifford* in this chapter derive from this source.

took place aboard the *Gifford* in the subsequent eight months. Not long after the burial, Veeder invited aboard a coterie of native women, including his mistress. This was not to be an overnight stay or even a week-long frolic but an extended visit in which the women were made partners in the whaling enterprise and at the same time awarded all the honors of visiting dignitaries.

Captain Veeder was a devoted host. He steered his ship close to shore so that his guests could hunt for seashells. He personally caught fresh fish for their dinners, and he made squirt guns for them to sport with on deck. Not entirely forgetful of the ship's mission, he and his mates assisted the women aloft and stood lookouts next to them. On October 28, 1871, the captain offered the women a bounty if they "raised" whales. They actually did so five days later, but the captain was so preoccupied with lovemaking that he did not bother to determine the species of whale sighted and he ordered no pursuit.

By the spring of 1872, Veeder's obsessions took a more dangerous turn. He had always had a liking for alcohol, but in the early stages of his debauchery he had relegated his consumption to week-long "drunks" on shore. Now he became intoxicated on shipboard and sought to navigate and command his ship while inebriated. On May 18, Veeder left orders for his crew to steer a particular course for the night. Had they followed his orders, the vessel would have been wrecked on a reef. Even so, "about half an hour after [the] watch got turned in," reported diarist Edward Kirwin, "all hands were called to save our lives as well as the ship."

It was not until May 19, 1872, that the crew of the *William Gifford* took matters into their own hands. The captain, fresh from beating his mistress, loaded two revolvers and came into the forecastle to look for her. The crew wrestled the shipmaster to the deck and tied him up. They then went aft and told the officers to sail to the American consul at Tahiti, "so that whoever was in the wrong would get punished." The sailors aboard the ship calculated that during the previous year the captain had had women aboard the ship for "210 days and the boat has been ashore 158 times."

The consul, one Dorence Atwater, found the sailors' testimony convincing. On June third he reported in a letter to the assistant secretary of state that he had removed Veeder as master of the whaling bark and had placed on board that vessel a "new master with instructions to proceed to San Francisco and there await orders from the agents." In reporting the fiasco of the whaleship to the federal government, Atwater tried to explain why the voyage had been such a disaster. One of the reasons he offered was that there was "no one on board save the Captain, who understood navigation sufficiently to navigate the ship."[93]

The pathetic and frightening story of Captain Richard Veeder is in many

---

[93] Dorence Atwater to the Assistant Secretary of State, 3 June 1872, Tahiti Despatches, National Archives, Washington, DC.

ways exceptional. Most shipmasters were too attentive to owners' interests and
their own employment prospects to risk ruining a voyage or a career in such a
way. Or they were mindful of legal consequences. Or they were better Victorians, too committed to self-restraint to have followed such a flagrantly "sinful"
path in the first place. Shipmasters managed their ships in very different ways,
but one thing that stands out in the social records of the whale fishery is the
degree to which most captains felt the constraints imposed by the distant
American shore.

The sea captain had a critical effect on the social character of a whaling voyage.
As a man of myriad postures and disciplinary styles, he might unite the ship as
a whole in pursuit of a common objective, or he might unite the forecastle crew
against him through cruelty. He could be an ally to a sailor, an adversary, or
both. The extent to which the forecastle crew forged a political community of
its own aboard ship depended a great deal on the peculiar and mercurial character of the captain. But it was also contingent on the internal dynamics of the
forecastle itself. Whether sailors set aside differences, found common purpose,
and took action together had as much to do with their own interests as it had
to do with immediate challenges from the men in command. The case of the
William Gifford is provocative in this regard. Why, we wonder, did not the men
of this ship stand up sooner to the obviously criminal Richard Veeder? The
reluctance or inability of the sailors to unite and protest nearly cost them their
lives. Were they unable to agree among themselves? Were some more cautious
about protest than others? Had some of them allied themselves to the men aft
and not to their shipmates forward? We cannot answer these questions for this
crew, but other sailors on different ships are more forthcoming. They describe
the social and political character of the forecastle, and they tell us about sailors'
collective will: how it was achieved, when it was used, and whether it worked.
It was before the mast, historians have said, that sailors in the merchant service
fostered a distinctive set of attitudes and forged a singular culture. It remains to
be seen whether men of the whaling forecastle faced common concerns and
found collective strength in similar ways.

5

Crossing the Line

Fraternity in the Forecastle

On the eve of a ship's departure, as outfitters rushed to fill out supplies, as sailors met in the forecastle for the first time, and as families steeled themselves for years of separation, whaling agents sat down at their writing desks to compose a letter to their captains. In this traditional exercise, they set the goals and terms of the voyage. They talked to shipmasters about their reputations, their responsibilities, and their futures. They spoke of the respect they had for these men and described how they trusted them to make wise decisions for their property. Without fail, they emphasized their great expectations for a full ship.[1]

The shipowners' letter – a show of power and authority at the highest level – would not be the only demonstration of its kind among the industry's professionals. Hardly had the last hills of home faded from view, in fact, when sea captains stood on the ship's quarterdeck to deliver an address of their own. They gathered their crews around them and set the terms and the tone of the voyage. "Men," the master began on the ship *Franklin*, "you have come to get a voyage have you not? I want you to come when you are called!" The *Atkins Adams* master sounded the same theme: "You've got to . . . go when you're told," he told his sailors. "You'll have plenty to eat and good provisions, as much as you can lay your sides to, so long as you don't eat to waste – you must learn the ropes quick."[2]

[1] See, for example, George Howland, Sr., to Captain E. A. Luce, 7 November 1837, "The Howland Collection," Mss 7, S–g 1, S–D, vols. 3 & 4 (letterbooks), WMLOD; Swift & Allen to Captains Blivin and Hamblin, 29 September 1859, "Swift & Allen Collection," Mss 5, S–g 3, S–F, S–s 2, Vols. 1–4 (letterbooks), WMLOD; Jonathan Bourne, Jr., to Captain James W. Munroe, 8 December 1854, "Jonathan Bourne, Jr., Business Records," Mss 18, S–G, S–s 2, vols. 1–3 (letterbooks), WMLOD.

[2] Diary of James Payne, *Franklin*, 5 September 1853, PPL; Diary of William Abbe, *Atkins Adams*, 11 October 1858, Log 485, WMLOD.

"There are," said the captain of the *Zone*, "on such a voyage as this, a great many hard thoughts and feelings; and some times we may seem rather over-bearing, but it will be all the better for you in the end, for I have been through it all, and know just what it is." But, glowered the *Mercury's* captain to his seamen, "you must have no quarreling, no striking each other. If there is any striking to be done, I am the one to do it . . . and, as long as you behave yourselves like *men*, you will get treated as such, but if you dont *by God* you will smell h+ll now you can go forward."[3]

Rituals of indoctrination and demands for obedience did not end with the captain's speech to the crew. Within a month or so of departure, veteran mariners before the mast sometimes staged a ceremony of their own. Close to the crossing of the equator, or the "line," as it was called, they initiated greenhands into a seafaring brotherhood. Symbolically separating neophytes from their shoreside kin, experienced mariners exhorted recruits to embrace their ship-mates as family and to bend themselves to the rules of the ship. They thus underscored the unity of the vessel as a whole. But more emphatically, they proclaimed the collective autonomy of the forecastle itself.

The ritual process of "crossing the line" was not smooth or painless. Greenhands balked and protested against being physically hazed, and they sometimes obstructed the ceremony entirely. Their noncompliance during this ceremony was emblematic of more sustained resistance. Throughout a voyage, men sometimes stood against the unifying force of the forecastle. But at other times, out of necessity, they became part of a powerful coalition before the mast. Brotherhood at sea on whaling ships was a tenuous reality, which linked sailors together and loosened the social bridge to shore.

I

On Tuesday, October 30, 1849, the ship *Hannibal* of New London approached the equatorial line. Sometime around seven o'clock that "pleasant" evening, sailors who had been working and relaxing on deck were startled by a cry from the masthead. "Sail ho," shouted the lookout. "Where away?" queried the officer on the deck. "Three points off the lee bow," the lookout responded. "What does she look like?" the officer asked. "A low black schooner," came the reply.[4]

At the report of the schooner, sixteen-year-old Nathaniel Morgan, who had been sitting on deck mending sails, was hastily ordered below along with other greenhands. Morgan knew well enough that his whaleship and its crew were in

[3] Diary of mate, *Zone*, 4 September 1855, KWM; Diary of Stephen Curtis, *Mercury*, c. May 1841, NBFPL.

[4] Diary of Nathaniel Saxton Morgan, *Hannibal*, 30 October 1849, MSM.

no serious danger, and that the schooner itself was a fiction, but he and his mates were nevertheless considerably anxious about the approach of the vessel. They suspected it carried none other than King Neptune, God of the Seas. As first-time sailors, they were likely to meet this important and intimidating personage, and they imagined that it would not be a painless introduction. In fact they heard him now, roughly addressing the *Hannibal*'s crew. "Ship ahoy! Are any of my children aboard?" "A few – come aboard," one of the sailors replied. Morgan could then make out on the deck above him the sound of "a tremendous trampling . . . amidst the rattling of chains and rigging." "Let me see some of my children," demanded the "old Veteran God of the Waters," and the ritual began.[5]

One by one, greenhands were blindfolded and taken on deck. Morgan was the fourth "child" to be released from the forecastle, led into the waist of the ship, and placed on a seat. Morgan could not see "Old Father Nep," but he could hear him rattling around him and he seemed in "most excellent spirits." He could also hear some unearthly instrumental music, which he later realized was produced by an old iron hoop drawn across an empty water cask. The young sailor was then put to the test and asked "a number of questions." He answered them, he said, "with all due respect." Then he was ordered to be shaved. "Accordingly," reported Morgan, "I was lathered with a patent chemical soap, compounded of Tar and slush and then most delightfully shaved, with this same identical musical iron hoop." At the words "Rinse him," Morgan was suddenly thrown backwards into a deck tub full of water. Soaking and dripping from his baptism, he was then informed that he was "initiated regularly" and could watch the subsequent ceremonies. When all of his fellow greenhands had been dunked and shaved and sworn in, "the whole crew and officers had a regular time of throwing buckets of water at each other." "Those who were not soaked through were scarce," reported Morgan, "and I think old Nep himself for once got as much salt water as he wanted."[6]

Whalers like the *Hannibal* were not the only ships that entertained King Neptune and served as the site for sailors' baptism.[7] The King of the Deep had been visiting all sorts of sailing ships for several hundred years, and evidence suggests that the ritual may have been more widely practiced in earlier centuries.[8]

---

[5] Ibid.   [6] Ibid.

[7] A number of folklorists, anthropologists, and historians have studied the Neptune ritual. Perhaps the most thorough analyses are Harry Miller Lydenberg's *Crossing the Line* (New York: New York Public Library, 1957) and Henning Henningsen's *Crossing the Equator* (Copenhagen: Munksgaard, 1961).

[8] Hawser Martingale (John S. Sleeper), *Tales of the Ocean* (Boston: J. S. Locke, 1840), p. 166, claims the Neptune ceremony was on the wane in the nineteenth century. The large numbers of greenhands on whalers may have made hazing on whalers increasingly problematic. Approximately one-fourth of the voyage records examined here speak of entertaining King Neptune on their ships and others mention the ceremony in general.

Edward Haskell, a merchant seaman aboard the *Tarquin* in 1862, drew this sketch of the King Neptune ceremony, also known as "crossing the line." An old salt has a forceful grip on a greenhand who has been lathered with tar. Once the hapless novice is shaved, he will be catapulted back into the water-filled barrel and baptized. King Neptune, at the lower right, congratulates a dripping initiate.

The Neptune visit that was staged on these vessels reflected the interests of its sailor perpetrators. It was an obvious rite of passage. Many of these young seamen negotiated between dependence and economic responsibility and were poised between social allegiance to parents and to a spouse and family of their own. They stood at a threshold in life, and Neptune spoke aptly to them.

The King of the Deep also addressed the concerns of men who hailed from working-class backgrounds, as many seamen did. He brought men together into a brotherhood and thus echoed the "mutualist" ethos of working men ashore.[9] And his message may have articulated and encouraged class antipathy. Published accounts of the ceremony suggest that seamen had to swear to social laws that contradicted bourgeois etiquette and claimed sailors' right to material well-being. They had to pledge to ride instead of to walk, to eat white bread instead of brown, and to swear to adulterous flirtation: "Will you pledge yourself," asked Neptune, "never to kiss the maid, when you can kiss the mistress, unless, indeed, you should like the maid best?"[10]

Although the God of the Sea spoke relevantly to laboring men who were young, he also spoke to the needs of men as sailors. He staged his visit during the ship's passage through the doldrums, that area of equatorial ocean known for its fickle winds and periodic calms. He thus provided seamen with important diversion at a frustrating stage of the voyage. Neptune also played a role as a social arbiter and attempted to unify mariners of various ranks and ages. Veteran sailors who served before the mast staged the ceremony and were permitted a brief but legitimate opportunity to take command over other sailors – the less fortunate greenhands. Although the subversion of "proper" order might come dangerously close to mutinous conduct – Morgan and the ship's officers had a grand time throwing water in each others' faces – the role reversal may have partly relieved the tensions between officers and crew and softened the hardships of shipboard hierarchy.[11]

The ceremony could have served the whole ship's company in other ways as well. (If it had not, the captain would surely have prohibited it.) On a number of ships, Neptune and his court quizzed initiates on nomenclature and facilitated the training of novices. They also strengthened bonds between officers and crew by underscoring the distinction between deepwater seamen like themselves and

---

[9] On the fraternalism of working-class culture, see Mary Ann Clawson, *Constructing Brotherhood: Class, Gender, and Fraternalism* (Princeton: Princeton University Press, 1989), p. 257.

[10] William M. Bryant, *The Old Sailor: The Thrilling Narrative of the Life and Adventures of Elias Hutchins* (Biddeford, ME: 1853), p. 9; Martingale, *Tales of the Ocean*, pp. 169–70. Marcus Rediker, *Between the Devil and the Deep Blue Sea: Merchant Seamen, Pirates and the Anglo-American Maritime World* (Cambridge, UK: Cambridge University Press, 1987), p. 166, has noted that the "rough talk" of eighteenth-century sailors expressed a similar defiance of bourgeois values.

[11] A general discussion of how initiation rites serve to resolve tensions in power hierarchies is set forth in Victor Turner's classic, *The Ritual Process* (Ithaca: Cornell University Press, 1969). For a more specific discussion relating to contemporary shipboard life, see Keith Richardson, "Polliwogs and Shellbacks: An Analysis of the Equator-Crossing Ritual," *Western Folklore* 36 (April 1977): 154–9. Whaling shipmasters, perhaps not wanting to walk the fine line between mutiny and the release of tension, sometimes forbade the ceremony. See, for instance, Diary of William Mitchell Stetson, *Arab*, 19 November 1853, Log 507, WMLOD; and Diary of anonymous mate, *Brunette*, 24 November 1842, PPL.

coastal sailors and between mariners and landsmen. Through the ministrations of Father Neptune, shipmasters were linked with the lowliest of greenhands in an elite community of ocean voyagers.

King Neptune thus reinforced the cohesion and distinction of an entire ship's company. In doing so, he underlined the cooperation that was necessary to the vessel's safe conduct and to the success of the whale hunt. But the God of the Sea emphasized more than the community of officers and sailors as a whole. He called on forecastle men to bind themselves together independently. During his ceremony, he relegated the ship's officers to peripheral or service roles as observers or as donors of refreshments. He also promoted familial solidarity before the mast. Neptune's sons became brothers to each other, and as siblings they were expected to provide social support and protection. On the *Samuel Robertson* in 1849, the "old fellow" told "his boys" that "they was initiated as sailors and could go to any part of the world without being molested." In another ceremony he asked sailors to "purchase a gallon of RUM for their shipmates the first time they should go on shore."[12]

King Neptune's lessons were not always happily learned. Many fledgling sailors were strong-armed into the seafaring brotherhood, and some of them fought back. On the *Hannibal*, greenhands gave Neptune a good dose of cold salt water. A novice seaman on the *Arab* in 1842 threw a copper funnel (the speaking trumpet) "into the face of his godship and cut his lip." He then did penance in a tub of water "until his divine anger was appeased.[13] Sailors who were not formally initiated into the maritime brotherhood expressed considerable relief at their escape from the rough trial, and greenhands who were in the numerical majority were sometimes able to persuade experienced mariners to forgo the event. Elias Trotter, a greenhand aboard the whaleship *Illinois*, described the forecastle of that ship as a scene of desperate anxiety as Neptune approached, with his fellow sailors "trying to form themselves into a body to resist the examination of Old Nep & others plotting means of escape by cowardly flight."[14]

The tensions revealed in the Neptune ceremony were frequently tinged with good humor and merriment, but they could also be quite serious. Beneath conflict-ridden ceremonies, and certainly behind those ritual events that never quite pulled together, ran real obstacles to forecastle unity. Men before the mast were separated from one another and allied to men aft by the lay system, the watch system, and the organization of boats' crews. They were also pulled apart by class, race, and ethnic divisions. As foremast hand Richard Boyenton remarked, whalemen were a "hetregenious mass of incongruity," and sailors

---

[12] Diary of George Coggeshall, *Samuel Robertson*, 23 October 1849, KWM; Bryant, *Old Sailor*, p. 9.

[13] Morgan diary, 30 October 1849; Diary of Samuel Chase, *Arab*, 12 November 1842, MeHS.

[14] Diary of Elias W. Trotter, *Illinois*, 22 August 1845, Log 1005 A & B, WMLOD.

This strikingly honest depiction of social relations at sea is a watercolor
sketch from the journal of the *Orray Taft*, painted sometime in the mid-
1860s. The three men are immobilized by wrist irons, a device commonly
used in ship discipline.

sometimes held tightly to the social prejudices they brought with them and that
divided them.[15]

Records speak particularly to conflicts – ranging from rancorous cursing to
jackknife stabbings – that had racial components.[16] William Allen, for one, was

[15] Diary of Richard Boyenton, *Bengal*, 30 May 1834, EI.
[16] Sailors on these ships call into question historical accounts that argue that whaling and
deepwater sail were distinctly free from racial discord. Michael Cohn and Michael Platzer,

awakened one night by a "most terrific engagement between a white man and a darkie" that went on until the bodies of the two men "were a perfect gore of sweat."[17] Later in his voyage, Allen reported that "a Nigger belonging to the Starboard watch did not turn out tonight when he was called . . . the consequence was that some of the men took a rope and made fast to both of his legs and [he] was hauled out of his bunk by 1/2 dozen men on deck half way up the gangway and then one of them poured a bucket of cold water on a part of his body which shall be nameless which made the old Darkle kick and holler gor' amighty."[18]

Harassment could be even more brutal. On January 15, 1838, on the way north at the end of a whaling voyage, sailors aboard the *Columbus* attacked one William Knobb, using the equatorial crossing ceremony as a cover. Journalkeeper Holden Willcox reported that "at 7 P.M. some of the crew undertook, according to costum to shave after Neptune fashion, a negro by the name of William Knobb. He therefore out of revenge or else having been instructed to do so stabbed C. B. Higgins in the back with a jack knife and cut Alexander Ware several cuts across the arm with the same Instrument. After this some of them gave him a sound drubbing, blacked his eye, swelled his lip &c." "Thanks to God," Willcox commented, "the voyage is nearly ended."[19] Race may not have been the driving force behind these hostilities, but the language of these diarists lets us know that it clearly played a part – a divisive part – in forecastle dynamics.[20]

The forecastle was also not free from overt class friction. Sailors from genteel Victorian homes often tried to elevate themselves above their co-workers and to link up with officers in the more bourgeois aftercabin. The most pretentious of these men were subject to rigorous teasing from their shipmates. Cyrus Chase, whose father was one of the "first men" in New York City and who had "never been used to any kind of work" claimed that he was the "butt of all hands on

in 1978, described whaling as a "truly integrated trade" in which "there were no records of fights conducted on racial lines." (Michael Cohn and Michael K. H. Platzer, *Black Men of the Sea* [New York: Dodd, Mead, 1978], p. 85.) Briton Cooper Busch, in his study of Cape Verde whalers, claims that Cohn and Platzer's argument "goes too far." (Briton Cooper Busch, "Cape Verdeans in the American Whaling and Sealing Industry, 1850–1900," *American Neptune* 45 [1985]: 113–15.) Jeffrey Bolster argues that the forecastle of the early antebellum period was a "unique workplace" where race might be "less of a determinant . . . than elsewhere." Later on in the nineteenth century, though, he says, "cross-racial camaraderie . . . seems to have eroded." See W. Jeffrey Bolster, " 'To Feel like a Man': Black Seamen in the Northern States, 1800–1860," *Journal of American History* 76 (March 1990): 1180, 1195.

[17] Diary of William A. Allen, *Samuel Robertson*, 29 August 1842, Log 1040, WMLOD.
[18] Ibid., 14 March 1843.
[19] Diary of Holden Willcox, *Columbus*, 15 January 1838, KWM.
[20] Racism and racial conflict were not specific to the forecastle, of course. For examples of racial hostility directed at stewards and cooks by officers and seamen, see Chapter 7.

A glimpse into a whaling forecastle on any given night or day would likely
reveal men engaged in the activities shown here: quarreling, smoking pipes,
and sleeping unsuccessfully. This engraving dates from the early 1800s.

board of [his] ship." A college-educated man who spoke five languages, Chase
asserted that he would "willingly dispense with all his accomplishments if he
could only be a good sailor man."[21]

James Haviland had similar difficulties. Haviland, who sailed aboard the
Baltic in 1856 to teach himself a "severe lesson," would have warmed the heart
of any seaman's reformer with his "steadiness" in port. He came back to the
ship sober, he saved his money instead of spending it in billiard rooms and
brothels, and in Honolulu he spent time at the "Sailor's Home." He was not
popular with his shipmates. The crew, he recorded on November 23, 1857,
"have quite an antipathy against me. They have endeavored while I was asleep
to haul me out of my bunk and take me on Deck."[22]

William Abbe, the Harvard alumnus who had joined a whaler to improve his
health in 1858, endured a lengthy comeuppance at the hands of his fellow
sailors. Abbe arrived on board the bark *Atkins Adams* with a dangerous social
agenda: he was eager to cement the social bond that he felt tied him to the
aftercabin. He made it a point, for example, always to bow to the shipmaster's
wife, and he sought to impress the mate with his knowledge of oceanic physics.
During his watch below he tried to distance himself from his shipmates by

[21] Diary of James Haviland, *Baltic*, 1 October 1857, PPL.   [22] Ibid., 23 November 1857.

submerging himself in works by Shakespeare and the Brontë sisters, reading behind his bunk's private curtain.[23]

While he was laughed at for his literary proclivities and chided for his penchant for privacy, it was Abbe's allegiance to the shipmaster's wife that opened him to special rebuke. Because he "wanted to be smart before the 'old woman,'" he refused to join some fellow sailors in a prank. Several members of the crew then chastised him by refusing to relieve him of duty at the ship's wheel.[24]

Social conflict, founded in race, class, or ethnic tensions and exacerbated by the competition between watches, was a common feature of forecastle life. It had a swift way of diminishing, though, in the face of adversity. When confronted with hardships perpetrated by a common "enemy" – the ship's captain – foremast hands were able to lay their animosities to rest. The experience of William Abbe, described above, provides a case in point. Abbe's fondness for the aftercabin and his distaste for his fellow hands lasted only as long as his good health. In late November 1858, after a month at sea, Abbe found himself "desperately hungry." Like the rest of his shipmates, he was forced to eat "sour & wormy" bread, sweetened by molasses that was "2–3 inches deep in cockroaches" and to drink water that made sailors "immediately . . . [throw it] up again." He found himself "desperately hungry," and while taking his trick at the wheel on November 23, he was lectured for "steering carelessly" because he was almost too faint and tired to stand.[25]

Ship fare improved somewhat over the next six months, and Abbe adjusted his palate so that he could even swallow cockroaches "without making a face." But Abbe's loyalty to the shipmaster had been broken. It was not the fact of shipboard hardship that had made him and his fellow seamen so miserable, he said; it was the "management" that seemed to exacerbate it. "Very little attention," he complained, "is paid to our comfort." As he encountered repeated abuse from the ship's officers, including being beaten in the face for "talking back" and being kicked by the shipmaster for leading a protest over inadequate rations, he found emotional strength in the collective misery and support around him. Together with his mates he planned his desertion from the vessel, and together they laughed over their unhappiness. Their noontime meal, especially, became both the subject of common anger and the target of shared humor. "Our duff this noon heavy & watery was literally filled with dirt & cockroaches," reported Abbe in early September 1859. "I didn't Eat a morsel of the filthy food but sat laughing at the discoveries the fellows made as they carefully sliced their duff. 'Hallo! heres a piece of old Thompsons hat cried Johnny – Heres a big worm – Look at these cockroaches – Ive bit a cockroach in two

[23] Abbe diary, 10–22 October 1858. [24] Ibid., 5 January 1859.
[25] Ibid., 22–5 November 1858.

– Lets make Thompson eat em when he comes below' – came from different empty mouths."[26]

As Abbe came to know and to identify with his shipmates, even those trappings of his Harvard education, used initially to isolate himself from fellow sailors, became communal property. He read his journal to members of his watch, and he shared Shakespeare with them. They in turn let him know what they thought of his "fancy" literary interests and informed him of their biases against the Bard. One sailor in particular, said Abbe, was strongly prejudiced against the playwright:

> Johnny Come Lately – who is the comedy of my forecastle life – keeps me awake for near an hour every night by his ludicrous fears – that I will turn out in my sleep and throttle him or do him some murdrous damage in a dream. He swears that I spout Shakespeare every night – & that once I stepped out onto my chest & struck him in the face – crying out "Come on Macduff & damned be he who first cries Hold out – Enough!" I get to laughing so at his absurd monkey gestures and talk – that I cant sleep – while Johnny shakes a coil of small rope – that he has provided he says to make me fast to a stanchion – the next time I attack him in a Shakespeare fit – He says I wont live long if I read Shakespeare much more – for its a bad book & will make me commit Suicide.[27]

As the slow shift in William Abbe demonstrates, rituals and prank playing alone could not enlist a foremast hand in the active cause of fraternity. Afterguard abuse was crucial in recasting "outsider" sailors into militant participants in the forecastle community. But the degree to which the afterguard helped to galvanize the forecastle was variable. Some shipmasters were honestly concerned about sailors' welfare and health, and treated seamen with respect and even generosity. Ships that lacked an antagonist may have left the forecastle – with all of its disparities – pretty much the way it was found. Many sea captains, however, economized at the expense of their crews, and sailors in their ships forged coalitions out of desperation.

The belief that shipmasters wilfully neglected sailors' health and safety sparked more anger and provoked more united protest from sailors than any other concern. Matters usually came to a head on this issue after six months or more into a voyage. During the earliest months under sail, when provisions aboard a ship were fullest and freshest and when shipmasters wanted to hang on to their men, the food served forward did not provoke many complaints. The cook delivered to the men before the mast regular and sometimes even ample supplies of salt pork or beef, biscuits, molasses, and coffee. He varied this staple diet with potatoes, beans, and rice, and whatever other vegetables remained on

---

[26] Ibid., 22–3 November 1858; 30 December 1858; 1 September 1859.
[27] Ibid., 13 August 1859.

hand. On Sundays, he often served duff, a boiled flour and lard pudding that was considered a special treat.

After a month at sea, it was not uncommon for a new mariner to offer generous comments about shipboard fare. Nathaniel Robinson, who was suffering from tuberculosis, wrote to his sister that he "was getting quite hearty" with sailor food. Asa Tobey, after three months at sea, asserted that "all hands are well on board & are doing well in the eating line plenty of fresh porpus & Duf to stuff under our shirts all fat as pigs & strong as lyons."[28]

But men who complimented the cook and the provisions and who flexed their healthy muscles tended to be relatively fresh from home. As voyages wore on, many forecastle hands began to tell tales, not of hardy stoutness, but of painful suffering and failing health. Debilitated from overexposure to the sun of the tropics, to the wind of the Capes, or to the cold of the Arctic, they reported themselves literally weatherbeaten. And ship fare, which at first seemed adequate, now seemed paltry in portions, and, being rotten or spoiled or worm-infested, it made men sick.

A few sailors looked at their smelly beef, their fouled water, and their cockroach-ridden bread and lay blame not on their captains but on the vicissitudes of ocean living. It was the sun, they said, that made the butter rancid and rotted the vegetables. It was the bad luck of headwinds and rainless weather that reduced the food supply. It was the moon, they suggested, that made their mates go blind.[29]

Whalemen who shrugged their shoulders at their miserable living and who attributed it to the peculiarities of their environment could sometimes read humor into their situations. Lewis Williams made a farcical "Bill of Fare" on Christmas day 1858 that included "pig, pork, hog, & swine boiled" and "water, both rain and river." And Edwin Pulver, of the *Columbus*, when forced to eat polar bear after his ship's provisions ran out, made the best of his misery by composing the following poem:

> I hope when we get more fresh meat
> It'll be of A kind that we can eat
> I care Not weather it be cow or hog
> Nor would I run from A well-cooked dog

[28] Nathaniel Robinson to Mary Robinson, 4–5 July 1843, PMS; Diary of Asa Tobey, *Houqua*, 1 November 1835, PPL.

[29] Diary of John Cleland, *Ceres*, 23 September 1836, MM; Diary of Abraham Head, *Lancer*, 22–4 December 1870, MSM; Diary of Lewis Williams, *Othello*, 12 October 1861, PMS. Several sailors remarked that they became "moonblind" when they slept on deck facing the moon. This may have been a symptom of scurvy. "Night blindness," noted Dr. Robert Hutchinson in 1907, "is a condition sometimes met with in patients suffering from scurvy who have also been much exposed to bright light, and may occur quite early in the disease." See "Scurvy," in William Osler, ed., *Modern Medicine: Its Theory and Practice* (Philadelphia: Lea Brothers, 1907), p. 899.

So here I vow likewise declare
I neer will Eat more polar bear.[30]

Other whalemen were not amused. They began to talk in unison about the human malevolence and incompetence that seemed to lie behind forecastle hunger and shipboard diseases. They readily targeted the cook as one agent of ill will and vented their frustrations on this galley worker with ridicule, pranks, and other forms of harassment. But they also cursed the captain. Convinced that he saved money for owners by scrimping on supplies for the men, they saw him living high off the hog while they starved. The smell of a fresh duck pie wafted forward to foremast hands of the *Atkins Adams* in June 1859, and they could just imagine the old man enjoying his "savoury" supper. "The spicy fragrance from the pastry came," said William Abbe, "deliciously to my nose." He imagined what such a meal might taste like. But his mind's picture corroded before him as the "filthy fingered" cook passed by with the salt junk for forecastle hands. John Joplin and his shipmates suffered a similar affront. In June 1848, their captain had "an extra blow out" for himself with wine and terrapin soup, while "Jack nasty face had duff and salt junk." Joplin's outrage increased a few weeks later when the shipmaster hung fresh pumpkins from the ship's spars. The pumpkins slowly rotted away, while the men went hungry. "If [the captain] only had brussels," Joplin snarled, "he would be one of the biggest damned hogs that Ever walked all that he thinks of is his own guts."[31]

Selfishness was bad enough, but planned cruelty was something else. Mariners were well aware of the fact that to force desertions from full whaleships, some captains became increasingly abusive. The more oil the captain of the *Henry Astor* got, "the worse he acted" complained one of his sailors, and he "done it on purpose to make his men run away!" The sailor had been determined "to stay by the ship but could not for the usage," and when he left he lost his outfit and $300 worth of earnings.[32]

The indifference or the malevolence of the master sometimes seemed to have no bounds, and harsh discipline, scant food, and verbal abuse prompted sailors to blame the captain for tragic events that in other contexts might have been ruled accidental. Charles Stedman described the time, on July 20, 1854, when a sailor missed his hold at the maintopmast crosstrees and fell 90 feet to the deck. Was this an accident? Stedman noted that the sailor was sick with fever and "ague" and was nevertheless sent aloft. William Allen described a similar event aboard his ship. Allen had gone to the masthead to look for whales at about four o'clock on the afternoon of November 25, 1842. Above him, on the

---

[30] Diary of Lewis Williams, *Gratitude*, 25 December 1858, PMS; Diary of Edwin Pulver, *Columbus*, 9 August 1852, PPL.

[31] Abbe diary, 7 June 1859; Diary of John Joplin, *Ann Parry*, 6, 17 June 1848, PMS.

[32] William Allen diary, 27 August 1842.

foretopgallant crosstrees, stood a sailor from New York state, named George Stevens. Allen had been aloft only half an hour when he was "startled by a short exclamation of distress" above him. He saw a body falling past him with "inconceivable velocity." As it passed Allen, it "was going feet downward in a slanting direction," then it lay "perfectly flat . . . and both arms stretched over [its] head as if trying to grasp something."[33]

It was in this position that Stevens (for it was he who had fallen) hit the water with "a terrible crash." A few seconds later, the sailor surfaced. His cap was off, and his "hair lay Floating on the water." The captain, meanwhile, had lowered a whaleboat, but by the time it reached the site of the accident, Stevens was 12 feet under the water. Soon afterward the captain braced the yards around, and Stevens was "left to his fate" in the ship's wake.[34]

Allen, hardened against the shipmaster after months of "despotism" and abuse, could not convince himself that Stevens's death was entirely an accident or that he was not recoverable. "Why did not the old man stop there all night," he asked himself, "making short tacks and keeping men on the lookout to see if he could hear anything of him and set lights? The man might only have been stunned." Together with his shipmates, he also began to wonder about the safety of the crosstrees themselves. How sound were they if Stevens had fallen so easily? The day after the sailor's death, men from the forecastle went aft and requested that the master send up his royal yards for the men "to lean on and hold on to instead of swinging about by the ships pitching and nothing to hold on to but 2 small ropes." In rough weather, they said, "a man is obliged to hold on so hard that in an hour his hands will be as bloodless and cold as a dead persons!" The captain's response only confirmed for the sailors his lack of concern. "No!" the shipmaster reportedly growled. "[I am] not going to have [my] mast lumbered up, there was no occasion for a man to tumble out of the crosstrees!!"[35]

Angered by poor usage and by apparent cruelty, some whalemen seethed and festered in silence. They secretly, independently, planned their desertions, or they plotted legal vengeance at the end of the voyage. A few men tried to appeal to the afterguard on their own. Sailor James Allen described October 20, 1870, as a "day long to be remembered" because a shipmate walked aft to the captain and told him the crew did not get enough to eat. He was turned away. "How we did laugh," Allen explained, "at the bold face of our hero."[36]

Many foremast hands decided that their best chances for better treatment lay not in private negotiations but in unified public demands. Suspending their differences, they planned together and approached the captain in a body. They showed him their stinking beef, their wormy and cockroach-laden duff, and

---

[33] Diary of Charles Stedman, *Mt. Wollaston*, 20 July 1854, NBFPL; William Allen diary, 25 November 1842.   [34] William Allen diary, 25 November 1842.

[35] Ibid.   [36] Diary of James Allen, *Alfred Gibbs*, 20 October 1870, PPL.

their paltry servings, and they described their fainting spells, their nausea, and the hollow pit in their stomachs. They told him they needed larger allowances of meat and bread, and more palatable food.

Sometimes sailors' appeals for better food and better treatment yielded little more than an angry reaction. Officers responded to protesters with blows to the head, knocks on the jaws, or hard, insulting kicks in the seat of the pants. Sailors might let the matter rest there, feeling that further protest would be futile. If a shipmaster had access to sympathetic port authorities or to alternative labor or if he felt no compunctions about beating or incarcerating or starving a man into submission, more strenuous confrontations would not work.

Sailors knew, too, that they might not be able to depend on assistance from the men in the steerage and that this might impede their chances for success. Mariners aboard the *John A. Parker* off the coast of Peru in 1853 "made a plot to strike for liberty," but when the mate called out "brace forward the fore yard," the braces were manned and the sailors were soon out at sea. Sailor James Bond explained that "the steerage gang backed out so it all fell through."[37]

The steerage company might have been reluctant to actively challenge the afterguard, especially in concert with foremast men, for several reasons. These mariners were closer to officers in rank than were foremast hands and they were paid more. According to William Abbe, they were "ambiguous – semi officers semi men . . . [with] all the insolence of command without any real authority." Many of them also had very different work to do from that of the men forward, and they thus had their own grievances. When steerage hands found reasons to complain, they sometimes acted independently.[38]

Whalemen of the forecastle might be limited in when and where they might stage protests and how much support they might expect from their shipmates further aft. That does not mean, though, that they did not sometimes succeed in their appeals. They occasionally strongarmed captains into providing better treatment, and even managed to punish some captains for past cruelty. Sailors' employment contracts, which they had signed at departure, had warned seamen to "obey the lawful commands of the Officers on board," and to "be good and faithful Seamen." Contracts warned sailors of the penalties they faced, such as demotion and loss of wages, if they were disobedient. And admiralty law prohibited protest as mutiny and prescribed cruel penalties for infractions. But angry seamen paid little attention to these directives and mandates. Faced with life-threatening hunger and repeated abuse, they stood up in solid defiance of the ship's formal authorities.

Whalemen not only used collective protests to bring about better food and living conditions; they also staged rebellions and obstructed shipboard labor.

[37] Diary of James Bond, *John A. Parker*, 14 July 1853, NBFPL.
[38] Abbe diary, 17 May 1859. See also Diary of William F. Maxfield, *Pioneer*, 4 August; 8 August 1871, Log 42, WMLOD.

Not infrequently they impeded the whale chase by slowing down boats or mishandling harpoons. Because shipmasters could not legally differentiate between hard luck and intentional clumsiness, it was difficult to formally chastise whalemen who repeatedly missed targets. Forced to make unhappy choices between losing whales and appeasing sailors, captains accommodated seamen. The master of the *Baltic*, said a seaman, built his crew a special wheelhouse for shelter because he knew that the crew had agreed not to apply much vigor to the chase. Men aboard the *Arab* in 1856 wanted to punish their shipmaster for stopping their dogwatch, and to do so they similarly lost some of their "fishiness." "The crew," reported William Stetson, "formerly appeared anxious to take oil [but] if the present state of things continues through the season our labor during the time will amount to but little. I for one sincerely hope that we may not take a drop of oil again until this season at least has passed away, and such a desire is not expressed by myself alone."[39] Such vindictiveness was founded not only on angry desperation but on the willingness of men to forfeit their own earnings. Perhaps hoping that the shipmaster had more money at stake in the season than they did individually and that he might also have a family to feed, these sailors counted on his additional vulnerability to make their point.[40]

Whalemen established some control over the ship not only by hampering the hunt but also by slowing the pace of sailing labor. On a few occasions, they used sea shanteys in this regard. These rhythmic songs helped sailors conduct some of the heaviest work on shipboard, such as weighing anchors or raising yards. The songs were organized by a shanteyman, a member of the crew who led the song and simultaneously unified men into a single body of muscle. The shanteyman was obviously helpful to the ship as a whole, but he could also be instrumental in managing work slowdowns. When he and his co-workers sang enthusiastically, one man was "as good as ten." When he slowed down the tempo or did not sing at all, sailors worked at a snail's pace. Mariners aboard the whaler *Atkins Adams* demonstrated their anger at the master, who had refused shore liberty, by leaving port "with slow & sullen strokes – without our accustomed shanties." Earlier in the voyage, these sailors chastised the master for complaining about their nighttime noise in similar fashion: "All of us hauled away in silence," the journalkeeper wrote, "but didn't gain a foot – purposely."[41]

[39] Haviland diary, 8 January 1857; Stetson diary, 17 May 1856.

[40] Very few ships in this sample (i.e., less than five) were the sites of violently desperate or vindictive behavior, such as setting fire to the ship, cutting the rigging, or attacking ships' officers. Only one full-fledged mutiny took place on these vessels, and that was described in Chapter 4.

[41] Richard Henry Dana, *Two Years Before the Mast* (1840; reprint, Los Angeles: Ward Ritchie Press, 1964, ed. John Haskell Kemble), p. 79; Abbe diary, 26 September 1859; 14 January 1859. Credit goes to Bob Webb for alerting me to the importance of shanteymen in shipboard labor negotiations. See also Robert Lloyd Webb, "Make Some Noise Boys! Deepwater Sea Chanteys," *Aquasphere* 16 (Winter 1982–3): 18–24.

Seamen were sometimes willing to extend their protests beyond slowdowns or false efforts, to outright refusals to work. Mutinous work stoppages were not uncommon on whalers, and they often led to arbitration that was handled quickly and informally. The circumstances of these collective refusals varied widely. Sailors on the ship *Rambler*, for example, decided not to obey orders that they considered life-threatening. In November 1855, this ship was cruising for whales in the South Pacific when she developed a leak. The captain, one Mr. Winchell, wanted to sail the ship to Chile so that he could ship his oil home. The sailors feared the vessel would not survive the distance. As boatsteerer James Payne reported, the crew gave the master a difficult choice:

> Mr. Thompson first officer gave orders to pump ship no one comeing to execute the order he came forward to learn the cause The men told him they had concluded not to pump any longer.... Capt. Winchell asked them why they had refused to pump ship they told him they [wanted] to go into port he told them he wanted to put the ship into Talcahuano where he could send his oil home and get the ship repaired but they refused to go there saying that the ship was not in a fit condition to go there he asked them if they would not pump ship they answered in the afirmative they told him that if he did not put her into the nearest port they should not pump anymore after some more words on the subject we came forward to give him a chance to make up his mind to let the ship sink or go into port at 8 1/2 o clock he gave orders to square the yards and keep the ship [south by west].

The seamen prevailed. In early December the ship reached the South Pacific port of Upolu in the Samoa Islands. She was condemned as unseaworthy, and the entire crew was discharged.[42]

Most power plays aboard whaling vessels involved less dramatic but similarly urgent issues over daily living and working conditions. Crew members aboard the bark *Covington* wanted more liberty days and, to add weight to their demand, refused to heave up the anchor. The master, after talking with them "to no purpose," increased their holiday. Likewise, the men aboard the *Smyrna* "refuse[d] to hoist anchor unless they [would] go to Angier." The captain "steered for Angier at 6 p.m." Five sailors went aft to the captain of the *George* in June 1840 and stated that they were willing to cruise for whales only if they had better meat & bread. He upgraded their meat allowance.[43]

Sailor collaboration and protest were so successful on some vessels that masters felt completely dominated by men before the mast. The crew of the *Reaper*, according to Captain Benjamin Neal, had the complete run of that ship. When foremast hands demanded that Neal go into port in the Indian Ocean, he

[42] Diary of James Payne, *Rambler*, 27 November, 12 December, 1855, PPL.
[43] Diary of Allen Newman, *Covington*, 11 February 1854, PPL; Diary of Charles W. Chase, *Smyrna*, 7 July 1855, Log 655, WMLOD; Diary of G.S.B., *George*, 28 June 1840, MSM.

headed in, even though he feared threatening weather. When the same men requested that he sleep ashore so that they could entertain women on board, he accommodated them. When they created "compleate hell on board, being beastly drunk," Neal looked the other way. "I have tryed to reason the case with them," he asserted, "but it is no use they have combined togeather."[44]

First mates, who were charged with noting mutinous or disrespectful conduct in logbooks, not infrequently looked on the acquiescence of shipmasters with disgust. "I am now sick boath in person and of this voyage," remarked Beriah Manchester, mate of the *Persia* in 1848, "for the chrew has the manageing of everything on board and have had for the most of the voyage and I am fairly worn out." He was no happier two weeks later when he noted that "the men work when they see fit and when thay dont think propper thay will not do a thing when told to do so."[45]

Richard Henry Dana, who studied the issue of sailor resistance at sea, might have found the behavior of these shipmasters hard to believe. Sea captains' authority and harsh discipline, he argued, left mariners with little recourse. "What is there for the sailors to do?" he remarked. "If they resist it is mutiny, if they succeed and take the vessel it is piracy. . . . If a sailor resists his commander, he resists the law, and piracy and submission are his only alternatives."[46] The evidence just described flies in the face of the opinions of this sailor-lawyer and suggests that he was perhaps more inclined to prescribe for sailors than to describe their actual conduct.

But Dana also derived his opinions from the merchant marine, where he sailed, as everyone knows, before the mast. Whalemen may have been more willing than their merchant counterparts to challenge the afterguard, and whaling masters may have been particularly vulnerable to power plays from the forecastle. As Captain Benjamin Neal explained, "to meet with success in a whaleing voyage it is of the utmost importance that all on board should haul together."[47] The lay system and the manner of the whale hunt meant that whaling masters were peculiarly dependent on the cooperation and hard efforts of their sailors. Whalemen also spent much of their time in isolated waters, where the lack of alternative labor and the distance from authorities may have

---

[44] Diary of Benjamin Neal, *Reaper*, 11 March 1839, EI.

[45] Diary of Beriah C. Manchester, *Persia*, 3 October 1848; 23 October 1848, Log 134, WMLOD.

[46] Dana, quoted in Bruce Nelson, *Workers on the Waterfront: Seamen, Longshoremen, and Unionism in the 1930's* (Urbana: University of Illinois Press, 1988), pp. 12–13. Concurring with Dana's assessment of the general futility of rebellious action is William Jeffrey Bolster, who examined insurrection on deepwater vessels (merchant and whaling) in the 1800s. Bolster also finds, though, that "as the period progressed, relative tolerance for minor disturbances slowly grew" and masters "grudgingly [provided] somewhat better supplies and working conditions." See William Jeffrey Bolster, "The Changing Nature of Maritime Insurrection," *Log of Mystic Seaport* 31 (Spring 1979): 15.

[47] Neal diary, 26 February 1839.

made officers more accommodating. Not least, the crews of whaling vessels were usually double or triple those of merchant vessels of the same size. When these men took issue with the afterguard en masse, they had numbers on their side.

In view of whalemen's coordinated protests and their sometimes successful efforts to flout afterguard authority, the nineteenth century cannot be considered, as it often has been, as the dark age of ship discipline. Historians of seafaring labor movements, who stress the accomplishments of seamen's unions in the 1900s, have tended to picture earlier American seamen – without specifying trade or decade – as "hapless." They also have dismissed sailor resistance as generally futile. Whalemen, if not all sailors to some extent, should be exempt from these generalizations.[48] When they laid aside their differences, found common cause, and acted in concert, whalemen became a powerful force, and they mitigated some of the worst realities of their deepwater worlds.

II

King Neptune and his surrogates had urged sailors before the mast to suppress their differences and to combine as a bold brotherhood. When faced with harsh discipline and abuse, whalemen did indeed shape a powerful and distinctively radical community before the mast. Neptune's followers had preached more than antiauthoritarianism, however. They had urged seamen to join together as a special set of friends – to form a fellowship of protection and social amity.

Fraternal togetherness was not something that came readily to men before the mast. In fact, to take a glance into the forecastle on almost any given occasion was to witness a multitude of personalities and diverse interests at work. The scene aboard the *Illinois* on a December night in 1846 may have been typical. Because the wind blew hard that evening, all but two of the nineteen sailors who lived in the forecastle were below. "There sits in yonder corner on buckets & molasses jugs around a chest," an observer wrote, "4 men immersed in the excitement of a game of 'all fours' – Behind them is an old Jack busy at modelling out a ship – In front of him is a fiddler in chrysalis, elbowing his way into music; near him is a game of 45 going on – In the center of the 'castle, several are engaged in the fore & aft dance with all the accompaniments of bad fiddling – overhead . . . lies one . . . reading his Bible. In yonder corner goes a scuffle, accompanied with horrid oaths & direst blackguardism & directly behind the scufflers lies another in his bunk reading D'Aubignes history of the reformation, perfectly unconscious of what is going on near or around him."[49]

---

[48] On "hapless" seamen see Joseph P. Goldberg, *The Maritime Story: A Study in Labor-Management Relations* (Cambridge, MA: Harvard University Press, 1958), pp. 11–15, as well as the discussion of Elmo Hohman in the introduction to this book. On sailors' lack of bargaining power before the 1900s, see Nelson, *Workers on the Waterfront*, pp. 12–14.

[49] Trotter diary, 10 December 1846.

Clouds of tobacco smoke curl around whalemen during the dogwatch. It
was during the early evening that sailors regularly enjoyed a short respite
from shipboard work. They often spent the time, usually from 6 to 8 P.M.,
spinning yarns, sharing gossip, singing, dancing, or taking a leisurely smoke.
(Robert Weir sketch.)

A similar scene unfolded aboard the *Eliza Adams* in January 1852, when John
Jones, a cook aboard that ship, went forward to "hear some music." Jones
reported that he "found the fidler playing the fourth of July Evans keeping time
with the bones, the blacksmith playing Juber on the banjo, Goss was playing
Bonapart crossing the alps on the fife, and Kimble was whistling Yankee Dodle
Do – the Portugues was singing a song of their own, and some of the rest was
singing old Dan Tucker is come to town."[50]

The whaling forecastle was obviously the setting for remarkable social
cacophony. But it could also be organized and unified. Men aboard whaleships
sometimes brought a community – often a geographic or age cohort – to sea
with them, and if they had a critical mass aboard they used it to bring hegemonic
cohesiveness to the rest of the group. Sailors aboard the *Esquimaux*, for example, all under the age of nineteen and almost all from Cape Cod, infected
other seamen with their enthusiasm for communal frolics. On one occasion they
gathered themselves and other sailors into a "military corps" and "went march-
ing about deck." At another time they convinced fellow sailors to participate
in group theatrics. They ran about the forecastle deck with cornmeal in their
mouths, "fluttering [and] striking their hands against their sides to resemble the

[50] Diary of John Jones, *Eliza Adams*, 7 January 1852, KWM.

wings of a fowl accompanied with crowing, cackling and gagging in imitation of a hen half shocked." Sailors aboard the *Atkins Adams*, similarly youthful, spent evenings after sunset organizing "whang o' doodle." This sport sent sailors chasing each other around the windlass, spanking buttocks. "We are kept," said a participant, "in a roar of laughter – at the contorted faces – and the rubbing with hands of the wounded parts."[51]

Foremast hands who hailed from the same social milieu ashore also gathered shipmates into less boisterous entertainments. American seamen organized crews into everything from shipboard "bees," to dramatic entertainments, to shooting matches. The crew of the *Lucy Ann* performed a complete musical revue for an audience of seagulls and offered their pelagic guests a variety of choral pieces and instrumental duets. This particular ensemble was interrupted by the afterguard, however. The production "wound up," journalkeeper John Martin explained, "by the Mate telling us if we did not quit making such a damned noise, he would heave a bucket of stinking water over us."[52]

Some of the most popular activities of the forecastle were those that actually accommodated, rather than submerged, social differences. One such was the "fore-and-after," a dance perfectly suited to the contrary interests of sailors. Mariners aboard the *Montreal*, who put together a fore-and-after in January 1852, had a "regular old fashioned 'break down,' 'drag out,' 'double shuffle,'" according to steward Washington Fosdick. It "made up in energy what it lacked in grace," he commented, "sounding very much like a locomotive at full speed over a rickety track where the rails were all loose." The fore-and-after, explained Fosdick, was highly suitable for the forecastle because "as each performer has his favorite step in which he excels, so as many steps can be seen at once in the same dance and to the same tune as there are colors to the rainbow all in rapid motion like the fly wheel of an engine. This then is the sailors favorite dance at sea, in which many an idle hour is whiled away of a calm, clear moonlight evening."[53]

As the case of the fore-and-after demonstrates, the forecastle worked as a social kaleidoscope of sorts, sustaining sailors' numerous interests. There were times, though, that demanded more single-minded unity from men before the mast. Just as true solidarity had emerged out of shared physical hardship, so sailors built an emotional community in the face of human loss. "Death is a solem occerence at anytime," explained James Payne of the *Monticello*, "[but] especially on board of a ship." "On shore," he continued, "when one is gone there [are] others to fill there places but on board of a ship you miss him." "There is," said Payne, "one less on a yard when you are on the yard you miss

[51] Diary of J. B. Hersey, *Esquimaux*, 25 April 1843; 19 May 1843, KWM; Abbe diary, 19 November 1858.   [52] Diary of John Martin, *Lucy Ann*, 16 February 1842, KWM.
[53] Diary of Washington Fosdick, (under pseudonym of Washington Foster), *Montreal*, 8 January 1852, Log 146, WMLOD.

his voice about deck you see him consigned to the deep without any friends to morne his departur."[54]

Whereas some sailors simply went on with the ship's business after the death of a sailor, others took comfort from a fraternal funeral of sorts. Richard Henry Dana, characterizing the ceremony in the 1850s as either a "law or universal custom," described events on his own ship. Shortly after a young English sailor was declared lost, the captain ordered the dead man's sea chest brought on deck, and presided as auctioneer over the chest's contents. Members of the ship's company bid on the various items until, as Dana recounted, "there was nothing left which could be called 'his.' "[55]

Aspects of this practice suggest that it had more symbolic than practical utility. Dana reported that foremast hands paid more for these secondhand clothes than they would have for new garments on shore. An auction on the *Cavalier*, held in honor of a sailor killed by a whale, saw things sell "very well for more then they cost." (One sign of this inflation was the sale of Dana's *The Seaman's Friend* for "one fifty.") When journalkeeper William Wilson took it upon himself to urge the men "not to bid foolishly," he was thought "hard." Yet the clothing and articles that were purchased so dearly would probably have been of little physical use to seamen. "Sailors," noted Dana, "have an unwillingness to wear a dead man's clothing during the same voyage, and they seldom do so unless they are in absolute want." On the *Cavalier*, sailors became "frightened" when the captain tried to use the marlinspike of the mariner who had been killed.[56]

If sailors paid high prices for items they could not use, they must have been purchasing something other than worn clothing. What they bought, it seems, were at least three things. First, they acquired a physical reminder of the drowned man, something that might mitigate the loss of his tangible self. Second, they made a contribution to the sailor's shore family, which would receive the auction proceeds from the captain. And third, they reinforced their own fraternity. By distributing the sailor's items among themselves, they reminded each other that this man's work and life at sea had been significant to all. Aboard the whaler *Ann Parry*, a French sailor died on August 7, 1847. His things were "divided up" and sailors then gathered to read his personal letters aloud.[57] What in other circumstances might be seen as a breach of propriety was in this context an act of collective grief. Those sailors who survived could take

[54] Diary of James Payne, *Monticello*, 12 July 1856, PPL. Had Payne been reading *Two Years Before the Mast?* His thoughts and words echo those of Dana on the same subject. See *Two Years*, p. 38. (And thanks to Jeff Bolster for pointing out the parallels.)

[55] Dana, *Two Years*, p. 39. Marcus Rediker is one of the first scholars to describe and assess the seaman's auction. See *Between the Devil*, pp. 197–8.

[56] Dana, *Two Years*, p. 39; Diary of William Wilson, *Cavalier*, 20 October 1849, MSM.

[57] Diary of Ezra Goodnough, *Ann Parry*, 7 August 1847, PMS.

comfort in the knowledge that if they, too, perished at sea away from the presence of family, they had another sort of kin nearby. In a brotherly gesture, fellow sailors would honor them with material sacrifice and shared affection.

Veteran sailors, acting as King Neptune and his court, but also playing themselves day-to-day, campaigned for community within the forecastle. They sought to forge among sailors a distinct alliance and a powerful political family. Although sailors resisted Neptune and his conglomerate if they could, many mariners extended their hands to their mates. Faced with physical pain or emotional suffering, they had no choice but to reach out to the different men around them.

Every man before the mast wrestled with these issues of community and conformity, and many wrote about them, but one whaleman outlined the nature of forecastle struggle with particular eloquence. Rather better known than the other sailors discussed so far, Herman Melville described in *Moby Dick* the distinctive qualities of seafaring fraternity and the tensions among individuals commited to this brotherhood and those opposed to it.

In his novel Melville introduces us to two champions of fraternity: Ishmael, a sailor from Manhattan, and Queequeg, a Pacific Islander. Ashore in New Bedford and Nantucket, while waiting to ship out, these two men discover the deep satisfaction of fellowship. Although they are men of different races and social backgrounds, they make a "cosy, loving pair."[58]

On shore these men are intimates. At sea, however, their love takes on a more communal tone and expands its social boundaries. Ishmael, the novel's narrator, occasionally rhapsodizes over the importance of fraternal love. A scene in which sailors gather together to liquefy crystallized whale oil finds Ishmael "squeezing my co-laborer's hands" and thinking "Oh! my dear fellow beings, why should we longer cherish any social acerbities, or know the slightest ill-humor or envy! Come; let us squeeze hands all round."[59]

It is Queequeg, perhaps more than any other sailor in the story, though, who recognizes the consummate value of collaboration. In contrast to Captain Ahab, with his self-centered search for the white whale, and to first mate Starbuck, who is ever-mindful of the ship's business and the owner's profits, Queequeg embraces a broader community. As he insists, it is a "mutual, joint-stock world, in all meridians." And it is, quite fittingly, Queequeg's life buoy that spins to the surface as the *Pequod* sinks, just in time to support the struggling Ishmael.[60]

Herman Melville was dealing with more than the whaling world in his emphasis on cooperation, self-abnegation, and racial fellowship. But Melville was a veteran foremast hand, and he grounded his work in historical fact. As an honest informant of deepwater culture, he knew what he was talking about when he suggested that brotherhood was the sailor's salvation.

[58] Herman Melville, *Moby Dick* (1851; reprint: New York: Norton, 1967), p. 54.
[59] Ibid., p. 348.   [60] Ibid., pp. 61, 470.

6

The Attack of the Daniel

Whalemen Ashore

Early in October 1825, a London whaleship came to anchor in the harbor of Lahaina on the island of Maui. Whalemen aboard that vessel, the *Daniel*, had just completed a cruise and looked forward to reveling ashore. We do not know if they found what other sailors sought: fresh food, strong liquor, and raucous entertainment. We do know, however, that one traditional reward of port stops – sex with native women – was denied them. A recent Hawaiian vice law, engineered by resident American missionaries, prohibited visiting sailors from entertaining women aboard their ships.[1]

The sailors of the *Daniel* did not take word of the new legislation easily. Through their master, one Captain Buckle, they appealed to an American missionary named William Richards to use his influence to have the taboo suspended. For four days they communicated to Richards their urgent and serious needs. On the fourth day, according to the missionary, the whalemen lost patience. They left the ship "in a body," and landed ashore in three boats "under a black flag." Armed with pistols and knives, they surrounded Richards's house. "Uttering the most abusive threats," they demanded the missionary's "life or his consent for females to go on board." Richards testified that he would have surrendered the former far more readily than the latter. He also reported, with no small amount of satisfaction, that Hawaiian natives had come to his rescue. Native chiefs had called out an armed force, which guarded his house until the day the *Daniel* sailed away.[2]

---

[1] Ernest Dodge, *Islands and Empires: Western Impact on the Pacific and East Asia* (Minneapolis: University of Minnesota Press, 1976), p. 82.

[2] *Hampshire Gazette*, Northampton, MA: 12 July 1826, col. 2.

The *Daniel*'s sailors were not the only mariners to object to the new proscription. In the wake of the edict, several missionaries reported that other
sailors had appeared at their doors demanding to know the reason they could
not see women. "We could only say," noted one, that "it is forbidden by the
word of God." The missionaries did not hesitate to name the agency at work in
the sinful mariners. "Satan," said one, "is making a great effort to oppose the
progress of our work."[3]

It is safe to assume that many of the historians who have studied whalemen
ashore, particularly in the Pacific, would have examined the record of the
Daniel, and nodded in assent. For they have largely depicted whalemen in port
as a crowd of single-minded men bent on defying or exasperating shore authorities, both native and introduced. "Whenever the carousing, drunken, dis-
ease-carrying, irrepressible whalers came ashore," noted one scholar in reference
to Tahiti, "they were the despair of the missionaries, a nuisance to the authorities,
and often the ruination of the Tahitians." Another historian, speaking of
Hawaii, argued recently that the "story of the settlement of the Hawaiian
Islands can be told in terms of the conflict between the lawless whalers and the
sanctimonious missionaries . . . whose objectives were diametrically opposed."[4]

The direct testimony of American whalemen calls into question this picture
of sailors ashore. It challenges especially the image of whalers as a socially
uniform group given to lawless rampaging. It indicates that many sailors ashore,
in the Pacific especially, were not invariably free to do damage as they pleased.
It also reveals that when whalemen were liberated or when they liberated
themselves through desertion, they often split apart and took pleasure individually
or in diverse ways. Many did, indeed, entertain themselves in "irrepressible"
carousing. Like their working counterparts on the American mainland, they
enjoyed recreational pleasures that included heavy drinking, gaming, and enthusiastic sexual sport.[5] But others took less noticeable pleasures in port as
independent tourists and spectators of foreign cultures, and some men quietly
sought the most pious activities they could find. The diverse ambitions of
whalemen, held in check at sea by distinctly difficult circumstances, were given
freer play in port. Foremast hands, especially, issued calls for political and

---

[3] Ibid.
[4] Dodge, *Islands and Empires*, p. 74; Richard Ellis, *Men and Whales* (New York: Knopf, 1991),
p. 163.
[5] On working class leisure, see Paul E. Johnson, *A Shopkeeper's Millennium: Society and Revivals
in Rochester, New York, 1815–1837* (New York: Hill & Wang, 1978), p. 56; Elliott Gorn,
The Manly Art: Bare-Knuckle Prize Fighting in America (Ithaca: Cornell University Press,
1986), pp. 133–5; Kathy Peiss, "'Charity Girls' and City Pleasures: Historical Notes on
Working-Class Sexuality, 1880–1920," in Kathy Peiss and Christina Simmons, eds., *Passion
and Power: Sexuality in History* (Philadelphia: Temple University Press, 1989), pp. 57–69.

social fraternity just as they did on shipboard, but it was only occasionally that they kept a tight brotherhood active and alive.[6]

Hundreds of American whaleships roamed the world's oceans in a given year, making landfalls at myriad remote places. Shipmasters stopped at uninhabited islands, where they procured wood, water, or fruit, and they also made regular, between-season stops at ports where they picked up mail, made repairs, shipped oil home, and reacquainted seamen with other members of their species. The Bay of Islands in New Zealand was one of the most popular places for recruitment in the Pacific, as were the Peruvian ports of Tumbes and Paita, the Chilean port of Talcahuano, and Tahiti and the Marquesas in the South Pacific. As the Arctic fishery grew, so did the strategic importance of Lahaina and Honolulu, so that by 1846, nearly 600 whaling vessels were visiting these Hawaiian ports each year.[7]

By the time a whaler finally touched shore, it carried men who were physically and emotionally desperate for change. Four to eight months of the same work, the same food, and the same conversation made sailors anxious for solid ground and diverse company. Even the sight and smell of land at a distance helped to strengthen sick and exhausted sailors. One mariner in the Indian Ocean sailed near an island with an orange grove, and he was transported with ecstasy. "To us whalers," he said, "who [have] smelt nothing more savory for eight months than whales blubber and bilge water" the "very air smelt delicious" so rich was it with "spices and other perfumes." Another man got his first scent of a land breeze and felt his coming release: "I feel," he wrote, "like a freed bird."[8]

Some of these men would be destined for disappointment. It was called *liberty*, but whalemen ashore were not necessarily independent. Shipowners and masters conspired to choose ports and to time visits not to suit sailors' inclinations but to suit their own concern for savings and expenditures. And, depending on the labor market ashore, as well as on the success and length of the voyage, masters might set men free or kick them out of a ship. Or they might watch them like hawks and hunt them like slaves.

The ease with which shipmasters recruited men ashore – and released sailors – varied. In places like the Azores or the Cape Verde Islands in the mid-1800s, native men eager to avoid domestic hardships provided a ready labor supply. Owners did not encourage runaways at this early stage of a voyage, but they were not devastated if a sailor left. According to one agent, captains could

[6] These conclusions rest on the experiences of whalemen globally but are most applicable to men who took shore leave in the Pacific. [7] Dodge, *Islands and Empires*, p. 81.

[8] Diary of George Blanchard, *America*, 30 May 1840, PC–TB; Diary of Robert Weir, *Clara Bell*, 6 June 1856, MSM.

always "get a plenty more."[9] As a consequence, whalemen were frequently given leave to roam these islands.

Giving liberty and hiring men on the western rim of South America seems to have been a different story. Talcahuano, complained one owner, "has got to be a very bad place [to find seamen]." Sailors testify that in Talcahuano and in the ports of Tumbes and Paita, they were either not allowed to land or were closely watched. "A spy on us all day," complained a sailor ashore in Paita, and other men echo his misery. On the coast of Peru, said whaleman James Bond, sailors "are hunted with bloodhounds worse than fugitive slaves in the states."[10] Ship's records at Paita and Tumbes, which are filled with receipts for "Watching and Hunting up men," give testimony to the comparison.[11]

Recruitment, discharge, and liberty in Pacific islands was altogether another matter. In the larger Pacific ports, replacement seamen – both Americans deserting from other ships and natives – may not have been hard to come by. In the Marquesas at midcentury, according to Greg Dening, the "turn-around of beachcombers was frequent and while the captains cursed the ease with which men ran away, they noted that it was also easy to pick up extra hands, whether beachcomber or Islander." In other ports of the Pacific, particularly at Tahiti and in New Zealand and Hawaii, whaling captains also shipped men with some ease. They were drawn to native labor because it was cheap: Pacific lays were regularly smaller than those paid to men originating in the United States. And these islanders, like those from the Atlantic islands, had reputations among American captains for bravery and boat skills.[12]

Being able to ship sailors easily was not the only thing masters considered in contemplating the nature of liberty for their sailors. Native labor may have

---

[9] Briton Cooper Busch, "Cape Verdeans in the American Whaling and Sealing Industry, 1850–1900," *American Neptune* 45 (1985): 104–16 (the quotation is from Busch, p. 108). On the same matter see Swift & Allen to Captain Josiah Chase, 5 December 1859, "Swift & Allen Collection, Mss 5, S–93 S–F, S–s 2, vols 1–4, WMLOD.

[10] Diary of William A. Abbe, *Atkins Adams*, 14 September 1859, Log 485, WMLOD; Diary of James Bond, *John A. Parker*, 13 July 1853, NBFPL.

[11] See, for example, receipt from Captain John S. Gardener to Pancho McGill, 14 June 1856, Paita, "Andrew Hicks Papers," Mss 105, S–C, S–s 9, Fol. 12, Bark *Champion*, WMLOD. Thanks to William Lofstrom for alerting me to the presence of such receipts.

[12] Greg Dening, *Islands and Beaches: Discourse on a Silent Land: Marquesas 1774–1880* (Honolulu: University Press of Hawaii, 1980), p. 245; Marion Diamond, "Queequeg's Crewmates: Pacific Islanders in the European Shipping Industry," *International Journal of Maritime History* 1 (December 1989): 123–40; Elmo P. Hohman, *The American Whaleman* (New York: Longmans, Green, 1928), p. 53. Hohman reports that in 1844 alone, 500 to 600 men from the Hawaiian islands shipped aboard American whalers.

(*Opposite*) This painting, which is one segment of a long panorama, depicts a whaleship at anchor in Huahine, in the Society Islands. At lower right, men in a whaleboat tow a raft of casks to shore for fresh water.

come cheap, but shipmasters had to pay fees to ship the men and in some cases to post bond for the return of natives to their home ports.[13] In addition, owners considered the training that they had already given to sailors for one to two years to be of some value. Owner Matthew Howland instructed a captain in his employ to try to entice his crews to stay "having trained them 2 years" and because, although he could (he said) find less expensive crews "a good crew is cheaper at high lays than a miserable crew is for nothing." Other owners similarly cautioned masters to avoid shipping substitutes and to hang on to their own men when the alternatives, although less costly, were "worthless."[14]

Far and away the biggest variable in a captain's attitude toward liberty, though, was whether his vessel was full or close to being full of oil. If a seaman from a full ship wanted liberty on shore, then more power to him: let him take as long as he wanted, preferably a permanent leave. He would then forfeit his earnings. But if that sailor was in debt, or if the ship needed more oil, then freedom ashore was no sure thing.

Shipmasters took several specific measures to control the extent and quality of liberty and to discourage desertion. They recruited for wood and water on uninhabited or uninhabitable islands, where men would not be tempted to leave. They paid money to native civilians and port officials to put sailors under surveillance. They punished the men they recaptured with some of the cruelest means they could conjure up. A sailor who ran from the *Albion* in the Bay of Islands in 1868 was forced to stand on top of the tryworks from November 29 until December 6, when he "was sent to his duty." On the *Baltic*, Captain Leonard Brownson heard that six men had deserted and declared that "if it costs him $5,000" he "will have them 6 men." He got four of them back and hauled them up onto the ship "with a rope around their Head." After getting them on deck, "the Old Mate flogged one of them and run long sticks [along] their legs and compelled them to sit in the Heat & Sun in this manner for 2 or 3 hours . . . they were not able to move hand nor foot."[15]

Such punitive tactics may have dampened the plans of many deserters, or such discipline may have backfired. One whaleman who ran away from the *Sea Fox* in 1870, while the ship was in Madagascar, got drunk when he was recaptured; his companion, though, "cut his throat to avoid being brought back." A sailor aboard another vessel was forcibly returned and "got out of his binds & cut off his right hand at the wrist." He would serve ships' officers no longer.[16]

[13] Hohman, *American Whaleman*, p. 53; p. 261.
[14] Matthew Howland to Captain Valentine Lewis, 16 January 1860, BLHU; Jonathan Bourne, Jr., to Captain F. A. Weld, 11 November 1859, "Jonathan Bourne, Jr., Business Records," Mss 18, S–G, S–s 2, vols. 1–3 (letterbooks), WMLOD.
[15] Diary of George Bowman, *Albion*, 29 November to 6 December 1868, PPL; Diary of James Haviland, *Baltic*, 5–6 November 1856, PPL.
[16] Diary of Anonymous, *Sea Fox*, 8 April 1870, KWM; Diary of Edwin Pulver, *Sea*, 28 December 1852, PPL.

This enlargement of an entry in the logbook of the whaleship *Elizabeth*, from 1849, indicates that four men, including the ship's cook, have made their escape. Apprehension of the deserters would yield a sizable prize for the time, and the reward money would be charged against the deserter's earnings.

Many sailors managed to get away permanently. The man who cut his throat recovered, then three months later "stold A boat And Ran Away."[17] Other disaffected mariners swam from their ships at night and were given refuge in the forecastles of other vessels. They braved pounding surf and predatory sharks and attempted to swim ashore. Most often, though, they disappeared while on shore, slipping into friendly native huts, eluding pursuers through crooked streets, or hiding out in nearby countryside.

Sometimes even fellow sailors did not notice a man's disappearance. John Martin and a shipmate nicknamed Lightning had been sent ashore to cut wood on Isla Grande, Brazil, in 1842. It was "awful hot" explained Martin, as the two worked to heave the cut wood down a hillside. Lightning, complaining of sickness and fatigue, lay down under a tree. "I did not miss him until the boat came to take us off to dinner," said Martin. "We never saw him afterwards." True to his nickname, Lightning had told Martin that he "intended to run away from the ship & get home by Rio Janerio." It is "supposed that some of the natives took care of him."[18]

[17] Anonymous diary, *Sea Fox*, July 1870.
[18] Diary of John Martin, *Lucy Ann*, 28 February 1842, KWM.

Ship logbooks attest to a high desertion rate among seamen, and any account of whaling and port life speaks of numerous beachcombers and deserters ashore in the Pacific at any one time. Elmo Hohman, who examined desertion in the records of fifteen whaling voyages between 1843 and 1862, claimed that there was a desertion rate among the original crews of around 29 percent.[19]

After emerging from hiding in hillsides or in beach villages, most men who jumped ship tried their luck on another whaler and eventually made their way home. A minority, however, were attracted to island life and stayed longer. And a few married into a native community and stayed permanently.[20] Unlike disaffected seamen a century earlier, though, deserting whalemen rarely resorted to primitive forms of rebellion such as piracy or organized to avenge their treatment aboard ship. If they had, the United States Navy would likely have been capable of hunting them down. One of the reasons for the reduction of Atlantic pirates in the early eighteenth century, according to Marcus Rediker, was an aggressive campaign by the Royal Navy to extinguish them.[21]

The high rate of individual desertion on whalers speaks not only to the oppressive circumstances aboard ship but also to the strength of sailors' separate interests. Community might suit an entire ship's company when faced with a whale or a gale at sea, and collectivism certainly characterized the forecastle at times, but many of the bonds that had linked men in shared interest dissolved ashore. Social differences that had been held in abeyance on shipboard were now given free rein, in fact, and conflicts between sailors that had been building for months waited for the space of dockside to be settled. Mariners who had been permitted liberty sprees often arrived back on board injured as a consequence of scuffles with fellow seamen. Men of the *Lucy Ann*, ashore in 1842, "settled the dispute they commenced some time ago." Portuguese sailors and Americans from the *Montreal* broke into a stabbing fight in New Zealand in 1852. And fifteen-year-old William Stetson, a cabin boy aboard the *Arab* in 1854, described relatively typical results from a port stay after visiting Talcahuano: "We have now had altogether eight days of liberty . . . the necessary and indispensable amount of broken heads, black eyes, bloody noses and otherwise damaged countenances has been given and received."[22]

[19] Hohman, *American Whaleman*, p. 64.

[20] Dening, *Islands and Beaches*, p. 130, claims that a small population of mariners "lived long lives" in the Marquesas.

[21] Marcus Rediker, *Between the Devil and the Deep Blue Sea: Merchant Seamen, Pirates, and the Anglo-American Maritime World 1700–1750* (Cambridge, UK: Cambridge University Press, 1989), p. 283.

[22] Diary of John Martin, *Lucy Ann*, 3 August 1842, KWM; Diary of Washington Fosdick, (under pseudonym Washington Foster), *Montreal*, 23 January 1852, Log 146, WMLOD; Diary of William Stetson, *Arab*, 18 February 1854, Log 507, WMLOD.

Sailors separated in peace as well as in conflict. They split off from shipmates
singly or in twos or threes to take their pleasure in different places. Some went
off to gamble or to play games at billiard halls or bowling alleys. Ravenous for
fresh fruit, others made their way to local markets, where they satiated their
hunger with oranges, bananas, peaches, plantains, and breadfuit. Still others
went as fast as they could into lush countryside and rambled through fruit
groves and green pastures. William Allen of the *Samuel Robertson* hiked up the
slopes of one of the Juan Fernandez Islands. The green hillsides of that Pacific
island looked to him "like enchantment," and he pictured for himself what it
might mean to stay. "It seemed," he said, "as though I could have lived there
alone forever. I could not step a step without seeing something pleasing.
Flowers . . . grew in wild luxurience; every hill and vale were covered with
peaches, figs, quinces, &c."[23]

Younger whalemen in particular had riotously good times riding on hired
horses and seem to have provided as much entertainment for natives as for
themselves. Men ashore on the island of Celebes were especially enthusiastic
horsemen and celebrated acrobats. Fifteen-year-old John Randall noted in his
journal in 1853 that when "Jack" rode, very often the horse slipped "between
his legs." However, he noted that "if we do not fall off more than once in a
quarter of a mile we make tolerable fair weather of it the Natives appear to be
amused at our sports & seem anxious to help us to the worst horses they can
find." Sixteen years later, American sailors were still falling off horses in the
Celebes. Only by this time, a diarist could make a laughable comparison between sailor equestrians and horsemen of the Civil War. "Any pleasant afternoon when there is a watch on shore," noted steward Silliman Ives, "one might
reasonably conclude that Sheridan's cavalry had come to town as these hair
brained sons of the sea dash up the street, and around the market place at
breakneck speed. . . . The horsemanship displayed on these occasions is truly
wonderful and the grand and lofty tumbling performed by some of these amateur
cavaliers would do credit to the best gymnast in the business." Not all port
town citizens found inept horsemen so delightful. In Honolulu in 1844, mariners
were fined $5.00 for "racing or swift riding" in streets or roads, a punishment
equal to that for fornication.[24]

Whalemen also took independent forays into local communities. As some of
America's first global tourists, they were not shy about poking and peering into
whatever interested them. Fanning out into towns and countrysides, they ab-
sorbed themselves in native cultures. Whaleman John States entered a room in
Talcahuano and found himself in the company of a dead child dressed up in

[23] Diary of William Allen, *Samuel Robertson*, 26 February 1842, Log 1040, WMLOD.
[24] Diary of John Randall, *Cleora*, 5 March 1853, Log 699, WMLOD; Diary of Silliman Ives,
under pseudonym Murphy McGuire, *Sunbeam*, 15 September 1869, Log 618, WMLOD;
Hohman, *American Whaleman*, pp. 112–13.

ribbons and propped up in a chair. He was in the middle of a family funeral, and after examining crucifixes and family pictures, he sauntered out. George Blanchard, ashore in Santa Catarina, Brazil, was fascinated by a Good Friday procession which featured "24 young virgins (supposed to be)" dressed as angels and "glittering with gold." His mouth agape, he followed the young women into a cathedral, where a Grand Mass was being performed. He found himself in a world he had never known before. He saw "gilded idols," and a "blaze of light from a thousand wax candles." And then he heard the church choir. Their voices were so harmonious and ethereal that Blanchard was enraptured. "I was so completely carried away that I didn't know where I was, nor what I was, but there I stood, with . . . stareing eyes."[25]

Unearthly-sounding songs and strange women captivated other whalemen. Sailors visiting Tahiti and Hawaii regularly walked into native huts, made themselves comfortable on family mats, and fell asleep. William Allen wandered around a Tahiti harbor until he was "heartyly tired." He then went into a hut and persuaded its inhabitants to sing. "I never heard such beautiful singers," he wrote. "Their voices are finer than ever I heard in any other place and though I could not understand one word of their language yet I could not hear them enough." They sang him to sleep. George Blanchard woke up in a native hut in Oahu with more than pleasant dreams. He lay on a mat to "get a snooze" and when he opened his eyes he found fleas on "every part of my carcass." By the time he hobbled back to his ship, his legs and face were swollen, and he was so sick he went off duty. On the potency of Hawaiian fleas he wrote the following poem:

> To be clawed by a Lion, or bit by a Bear,
> Or snatched by an Eagle, aloft by the hair,
> Or crushed by an Elephant's huge proboscis
> Or sucked by a whale, down his open fauces,
> Or hugged to death by an Anaconda,
> Would *not* be a matter of special wonder;
> But thus to be vanquished, and left to bleed,
> By the *dirk* of a *Flea* tis strange indeed.[26]

Some whalemen, then, went ashore to tour and others to satiate their hunger. Some found rollicking adventure on the back of a horse, and others took their pleasure in billiard halls or tenpin alleys. Thus seamen gratified their different social wants. But while they all packed as much pleasure into their liberty as they could, sailors took two distinctly different paths to shoreside entertainment. The majority of whalemen went into port to drink and to become, as

---

[25] Diary of John States, *William Wirt*, 20 March 1846, MSM; Diary of George Blanchard, *Pantheon*, 26 February 1845, PC–TB.

[26] Allen diary, 21 October 1842; Diary of George Blanchard, *Solomon Saltus*, 6 October 1847, PC–TB.

several men put it, "gloriously drunk." And many went to town to have as
much sex as they could afford. As one blunt shipmaster put the matter: after
a visit on shore, his sailors were "fucked to death."[27] But other whalemen,
particularly those with Victorian sensibilities, turned their backs on such alarm-
ing activities and sought soothing recreation in reading rooms, in chapels, or
among missionaries. Their very different behavior ashore cautions us against
attributing single-minded interests to seamen in port.

Thirsty sailors looking for a blowout found a variety of intoxicants awaiting
them in whaling ports. They were offered aguardiente in Tumbes and rum in
Mahe. In Lahaina they tried tobacco beer, a concoction so strong that several
glasses would reduce a man to a state where he wouldn't "know his own name."
In the Cape Verde Islands sellers shouted "Brandy John! Brandy No Want
Brandy John, eh? Very good brandy John!" Sailors ashore bought liquor from
native sellers, in rum shops, and in groggeries. In New South Wales in 1842,
mariners could purchase their grog from a Mr. Sherritt, a preacher, who, when
he wasn't offering alchohol for sale, delivered sermons at church. Thus seamen
who had money to spend and who found liquor to their liking became, in the
words of one whaleman, "kind of loony."[28] When sailors went ashore sober and
had to be hoisted back aboard and when drunk men thronged the streets of
Pacific ports, reformers and missionaries recoiled in disgust. But they also got
to work. They tried to close ports to traffic in liquor in the mid-nineteenth
century, and they succeeded, off and on, in places like Maui. But the attempt
to turn the whole world's ports into dry towns was a lost cause.

Not uncommonly, drinking went hand in hand with a frolic among the
"ladies." Sailors visiting the Seychelles in 1847 dressed themselves in shore
togs, took pieces of calico in hand, and had "a great time amongst the women."
Seamen aboard the *Lucy Ann* had such a sweet time with Cape Verde women
that they "swore they could live there all the days of their lives & a spell longer."
When the time came for them to depart, the "tears were running out of their
eyes." They consoled themselves by remembering that their ship was nearly out
of wood, and they would have to visit a port and its inhabitants again soon.[29]

Women entertained whalemen ashore in their brothels or rooming houses.
They also accommodated sailors who remained aboard their vessels. In less-
populated harbors, they made "house" calls. The *Bengal* had barely let down
her anchor in the Marquesas in 1833 before scores of women, carrying clothing

[27] Diary of William Wilson, *Cavalier*, 9 March 1850, MSM.
[28] Stetson diary, 9 December 1854; Diary of Ezra Goodnough, *Ann Parry*, 5 July 1847, PMS;
Diary of James Haviland, *Baltic*, 31 October 1856, PPL; Diary of Dan Whitfield, *Dr.
Franklin*, 12 November 1856, PC; Martin diary, 31 July 1842; Goodnough diary, 1 March
1847.
[29] Goodnough diary, 1 March 1847; Martin diary, 17 January 1842.

In the popular press, the sailor ashore was synonymous with profligacy, sexual excess, and inebriation. Even the sailors' horses in this nineteenth-century illustration seem to stumble drunkenly.

with one hand and propelling themselves with the other, swam out to the ship. Once aboard, they adjusted their dress before they began to bargain with sailors over sex. Ten years later, sailors in the same harbor met "40 or 50 girls" willing to "grant any favor for from a head of tobacco up to a thin shirt or 50 cents." One of these seamen made the statistical claim that "2/3 of our whaleships when they cruise round these or any other islands . . . run in to the land at night send 2 boats or 3 on shore and fetch off girls 1 to a man fore and aft cabin boy and all included."[30]

[30] Diary of George Russell, *Bengal*, 4 March 1833, EI; Allen diary, 9 November 1843.

Kicking up her heels to the obvious interest of male company, this dancer animates the pages of Captain Frederick Smith's journal, kept aboard *Petrel* in 1873.

Women stayed aboard a ship sometimes for days, weeks, and, if whaling was possible nearby or if the shipmaster was detained on shore, for a month or more. The *Dr. Franklin*, which visited St. Helena in April 1859, became, according to one of its crew, a "perfect floating brothel." The captain periodically visited the ship but did not interfere with sailors' liaisons with women. In fact, according to journalkeeper Dan Whitfield, he "appeared to be pretty well acquainted" with all the women himself.[31]

[31] Whitfield diary, 8, 11 April 1859.

Sailors on the *Dr. Franklin* entertained their visitors from shore in fine style. Under the influence of free-flowing brandy, they staged an elaborate, week-long house party. It started with dancing – reels, jigs, polkas, waltzes, and cotillions – but when the "Fidler [was] too drunk to Play any More," they switched over to song. Sentimental favorites such as the "Dark Eyed Sailor" and "The Time is Coming that We Must Part" soon "had the women in tears." But these whalemen were nothing if not gallant, and according to Whitfield, "those of the Male Gender who were in A condition to help" their visitors escorted them to bed. The mate, however, had stopped playing the polite host. Fully inebriated and unconscious, he was "sprawled out on the Cabin Floor with Nothing but his shirt on."[32]

And so it went on the *Dr. Franklin*, "larking, joking & Love Making." On April 13 the whaleship's crew sought a wider audience for their performances. In the middle of the afternoon they stripped themselves of clothes, jumped into the water, and performed unspecified "diferent feats" for their women, who watched from the ship's rail. They then commandeered a whaleboat and rowed it around the harbor. They stopped aboard the whaler *Keoka*, went directly to the vessel's head, and positioned themselves strategically. They pointed their "Bare posteriors" toward a neighboring whaleship [the *Leander*] and shouted out "Look and Weep!" They then "let fly a full charge from their double batterys." The mate, still "dead drunk," missed the fun. The second mate was conscious enough to sing out a salute to the revelers, however. "Go it Gals," he shouted, "Wilcox pays for All!" Henry Wilcox was one of the owners of the ship.[33]

The curtain may have closed on the *Dr. Franklin*'s seamen at a particularly happy point in their spree. Several weeks later, these sailors may have been singing less jovial songs. Syphilitic diseases hit many of these whaling crews with a vengeance, and the diseases themselves, or the toxic mercury treatments for them, sometimes left mariners shadows of their former selves. Among the many crews afflicted with sexually transmitted diseases, for instance, were ten men of the *Montreal*, who were so sick they could not man whaleboats. More than half of the sailors aboard the *Cavalier* were similarly infected with syphilis, with "blind bulboes abundant." The mate was severely diseased, and according to a foremast hand, he was "but a shell, a mere wreck."[34]

Venereal infections were not the only thing that could dampen the spirits of carousing mariners. Observing at a safe and anxious distance the frolics of men

[32] Ibid., 12 April 1859.   [33] Ibid., 13 April 1859.

[34] On the treatment of venereal diseases, see Charles E. Rosenberg, "The Therapeutic Revolution: Medicine, Meaning, and Social Change in 19th-Century America," in *Sickness & Health in America*, Judith Walzer Leavitt and Ronald L. Numbers, eds. (Madison: University of Wisconsin Press, 1985), p. 42; Fosdick diary, 3 November 1851; Wilson diary, 27 March 1850.

like those on the *Dr. Franklin* were a few horrified fellow sailors, along with their social soul mates, American missionaries. Syphilis, these individuals suggested, amounted to nothing more than "just deserts."

Missionaries devoted no small amount of energy to eradicating the scourges that whalemen seemed to spread around the world. And they had been at their labors for some time. English missionaries had begun to arrive in the Pacific regions in the 1790s, and within two decades American reformers had sailed into the area. The American Board of Commissioners for Foreign Missions (ABCFM) was founded in Boston in 1810, and up until the Civil War it sent more workers abroad than any other agency. Driven by a belief in universal responsibility and by the conviction that all the world's people might be saved, this agency and its representatives endeavored to bring about no small share of the world's redemption.[35]

Early successes must have convinced American missionaries that they were guardians and proponents of a true faith. By 1837, ninety missionaries labored in Hawaii to train and convert natives, and they achieved what one historian has called "spectacular success." To a society "already in the process of overthrowing its ancient idols," says William Hutchinson, "God had sent the bright-eyed representatives of a republican experiment that thought of itself as the new order of the ages." Hawaiian royalty and leadership helped collaborate with missionaries in their evangelical efforts, and in the late 1830s and early 1840s there were massive religious revivals. By midcentury, Hawaii was used as a base for missionary activity for far-flung efforts in the Marshall, Gilbert, and Caroline Islands.[36]

Yet the story of American missionary activity in the Pacific is not one of smooth conversions. American efforts in the Marquesas, which began and ended in the early 1830s, were generally regarded as failures, and even in Hawaii, the efforts to Christianize the population were fraught with conflict and stubborn resistance. The problem, as missionaries put it, was not only that natives were sometimes reluctant to conform to the moral culture of Western Christianity but also that sailors and traders undermined their good works. Missionaries were dependent on sailing traffic and on the goods that ships brought, but they felt that they could have done without the seamen themselves. In the Marquesas, for example, missionaries left after only nine months because, in part, "beachcombers and sailors . . . had discredited their missionary witness." Vessels that had called at the islands during their stay had left "the most obscene and blasphemous words in the language." They also left women with venereal diseases. According to one scholar, the missionaries had

[35] Dodge, *Islands and Empires*, p. 118; William R. Hutchinson, *Errand to the World: American Protestant Thought and Foreign Missions* (Chicago: University of Chicago Press, 1987), pp. 45-7.

[36] Hutchinson, *Errand to the World*, p. 62; Dodge, *Islands and Empires*, p. 124.

initially tolerated ships' visits, for they hoped that islanders would become dependent on ships' goods and would be willing to work in a disciplined and "civilized" way for more of the same. But the plan to draw these natives into Christian capitalism obviously backfired.[37]

In Hawaii missionaries went about reforming both natives and sailors with considerable aggression. To limit the influence of the "wicked men" who opposed the "progress of the gospel," the missionaries began in the 1820s to get involved in the business of lawmaking. With the cooperation of native authorities, they pushed through edicts against prostitution, gambling, drinking, and breaking the Sabbath. Some among them, including the ABCFM's senior secretary, insisted that missionaries should serve more as a model of Christian behavior than as an active civilizing influence, but such cautionary remarks did not find many attentive listeners.[38]

The conflicts between American missionaries and American whalemen that built up in the antebellum period, particularly in Hawaiian ports, represented more than disagreements over salvation. Many whalemen correctly identified the class component in the conflict. "Our Sandwich Island mishionarys," remarked John Jones, aboard the *Eliza Adams* in 1852, ". . . are always ready to take what small change a poor Devil has in his breeches pockets on Liberty Days and in return for it will tel you that you will certinly go to hell for catching whales on Sunday to get that same money they are so wiling to take. O, consistency where art thou?" George Blanchard was similarly bitter: "As for the missionary operation I don't think but little of [it] – the fact is this: take them upon the whole they are too infernal lazy to get a living at home and so they get appointed missionary to some beautiful island of the Pacific [where they sit] laying up the dollars. Oh! it makes me feel ill all over to hear them at Home tell of how the poor Missionaries suffer – and indeed they do suffer – but it is with the Gout occasioned by lazyness and high living."[39]

Missionary and whaler disagreements may also have been characterized by gender conflict. Ministers, as Ann Douglas has pointed out, embraced behaviors and affects, such as piety and sympathy, that many whalemen considered feminine, and they often inhabited the same space women did – "the Sunday School, the parlor and the library."[40] Many whalemen, on the other hand, expressed a very different sort of masculinity and operated in different social space, involving,

---

[37] Dening, *Islands and Beaches*, pp. 182–4.

[38] Rufus Anderson, *History of the Sandwich Island Mission* (Boston: Congregational Publishing Society, 1870), pp. 64–5; Hutchinson, *Errand to the World*, pp. 70–90; Dodge, *Islands and Empires*, p. 126.

[39] Diary of John Jones, *Eliza Adams*, 19 January 1852, KWM; Blanchard diary, *Solomon Saltus*, 15 November 1847.

[40] Ann Douglas, *The Feminization of American Culture* (New York: Avon Books, 1977), pp. 48–9.

as we have seen, aggressive conquest in very undomestic settings. And while masculine ideals for ministers stressed self-control above all else, masculine conventions for many whalemen centered on relatively free expression.[41]

Relationships between the two groups might also have been more amicable had it not been for their mutual urgency. When whalemen landed on beaches and sent down anchors in foreign ports, they felt time constraints that compelled them to act with special intensity. Their visits to brothels, the sex they had on their ships, and the liquor they drank served as isolated points of rejuvenation between months of enforced abstinence. When they arrived on shore on liberty and found groggeries closed, carousing forbidden, and, as one sailor put it, that "the Missionarys has got glory pumped into the natives good and at both ends," their frustration was extreme.[42] Missionaries felt time constraints of their own. Not only was their own redemption tied to the degree of their struggle, but the salvation of the world at large, they felt, depended on their efforts to convert people to Christianity.

Given these tensions, it seems strange that missionaries and whalemen did not enter into more violent conflict. Yet the attack of the *Daniel*, the armed confrontation that began this chapter, appears to have been an unusual event. And the American missionary involved in the *Daniel*'s case remarked that, unlike the English, "American ships do not molest him."[43]

Determining why or if British whalemen were particularly prone to armed protest is beyond the scope of this study. There are, nevertheless, at least two reasons American whalemen may not have stood up to port authorities, including missionaries. The first has to do with the forces of law and order that were arrayed against them. The second centers on the heterogeneity of the sailors themselves. American whalemen were not uniformly antipathetic to missionary interests.

Sailors who contemplated militant protests knew that they faced an uphill battle. In the Hawaiian islands, for example, the police force was strong, active, and well connected with the United States' missionary establishment. At midcentury, a military company of foreign residents patrolled the streets and arrested boisterous sailors. A native militia of 200 to 300 stood ready to assist Governor Kekuanaoa. Finally, the governor's own soldiers could be pulled in to squelch any disturbance, and the governor himself was not hesitant about establishing curfews and about sending sailors who did not return to their ships after dark into the local lockup.[44]

Whalemen faced not only disciplinary authorities ashore but also the more

[41] Discussions of working class sexuality can be found in Peiss, "'Charity Girls' and City Pleasures," Peiss and Simmons, *Passion and Power*, pp. 57–69.
[42] Diary of Thomas Morrison, *Avola*, 13 March 1873, KWM.
[43] *Hampshire Gazette*, 12 July 1826, col. 2.
[44] Alexander, *Hawaiian People*, p. 274; *The Friend*, November 1852.

familiar restraining influence of their own captains. Unlike the *Daniel*'s captain, who sailed with a mistress and expected the same for his sailors ashore, many American shipmasters were fully sympathetic to the moral principles of missionaries and to their reform efforts. "We do not any of us like to go to Oahu," wrote several whaling captains to the governor of Maui, "because bad men sell rum to our seamen. We like your island, because you have a good law preventing the sale of this poison."[45] Captains like these, when confronted with ports where grog shops were licensed, endeavored to keep their men well offshore for much of their time in port, and they threatened or bribed seamen to stay sober. They also used liberty for disciplinary leverage over mariners. They promised extra days off for sober or restrained behavior on shore and for prompt return to the ship. They subjected crews to lengthy lectures on obligations and on the desirability of a visit to a chapel or a bethel. Finally, they attempted to control the extent of liberty with their management of liberty money, rarely allocating more than $1.00 for a sailor's spree.[46]

The ability or the desire to protest, though, had to come largely from the collective will of sailors. And here, it seems, lay the biggest obstacle. The collaboration that made forecastle resistance possible at sea was far more problematic in the wide-open social arena of a foreign port. In the confines of the forecastle, under the inescapable authority of the shipmaster, men who went to the sea with little in common were forced, for survival's sake, to come to terms with one another. But ashore, not only did ships' crews split apart physically but they also recognized no single adversary. The missionaries who antagonized some sailors represented welcome company to others. Not a few whalemen visited seamen's libraries and bethels and practiced temperance ashore. It would be difficult to marshall these men against missionaries in the same way that they had been organized to confront a man who starved them.

Victorian seamen in particular moved themselves away from their mates when they went into port. Their diaries, which had at sea indicated a coming to terms with men of other backgrounds, now registered disgust at shipmates who "let their basest animal passions weigh unchecked."[47] These upstanding men had come to sea to strengthen their self-control, not abandon it, and as they witnessed breach after breach of propriety ashore, they were reminded of the differences between them and their fellow sailors.

Cross-cultural and interracial relationships, too, which had been at times a necessary part of forecastle life, were now something that these outsiders sought

---

[45] Cummins E. Speakman, Jr., *Mowee: An Informal History of the Hawaiian Island* (Salem: Peabody Museum of Salem, 1978), p. 101. No date is given for this letter, but the context suggests the 1840s.

[46] Stetson diary, 18 February 1854; Diary of James Haviland, *Baltic*, 26–30 October 1856, PPL.

[47] Diary of Orson Shattuck (under pseudonym of Charles Perkins), *Frances*, 5 January 1852, Log 994, WMLOD.

to avoid. They relished the personal freedom of liberty days as much as the next sailor, but they were frequently displeased with the inevitable encounters with "Otherness." There were stopping places, to be sure, that did not jar their sensibilities. The Galapagos Islands, for example, offered Victorian whalemen nothing more offensive than land crabs and marine iguanas, and activities there rarely impinged on moral conventions. The islands of Juan Fernandez, too, which were the setting of Alexander Selkirk's solitary life and the inspiration for Defoe's *Robinson Crusoe* (published in 1719), had strong bourgeois connections. And St. Helena, the location of the orgy on the *Dr. Franklin*, offered men with other tastes the site of Napoleon's exile and death.[48]

But most anchorages brought Victorian mariners face to face with "grotesque" sights and offensively different cultures. Frequently, as they toured through the streets and paths of foreign ports, they were accosted by offers to partake in "sin." Richard Hixson said he could barely take a step through the village of Maui in 1834 without hearing "Owhynee? Owhynee? that is do you want a gal?" Orson Shattuck "had scarcely put [his] foot on shore" in New Zealand before a native man began bargaining with him for his daughter. Shattuck "refused with an expression of disgust." Yet the dealer was persistent and lowered his price. "I then spent about an hour," reported Shattuck, "in trying to make him understand but at last I had to give it up in despair for I could not make him perceive why it [was] wrong and as a last resort to get rid of him I told him that I was married." This apparently satisfied the New Zealander, although he commented that Shattuck was "the first white man that he had ever seen who was married."[49]

Whalemen who drew back in horror at the thought of sexual contact with native women grounded their feelings in a number of social concerns. Orson Shattuck wanted to match his purity to what he imagined was the continence of women at home. He could not bear to "pollute himself" with "creatures" because, in good conscience, he could never later embrace his virginal sisters. Furthermore, a former sweetheart, a woman who was "all purity and modesty," loomed large in his mind.[50]

Attitudes towards sexual liaison were, not surprisingly, infused with race prejudice. Women whose faces conformed most closely to white, Anglo-Saxon

[48] See Daniel Defoe, *The Life and Adventures of Robinson Crusoe, &C.*, reprinted in a Norton Critical Edition, Michael Shinagel, ed. (New York: Norton, 1975); Felix Markham, *Napoleon* (New York: Mentor, 1963). Napoleon was sent to St. Helena in 1815 and died there in 1821.

[49] Diary of Richard Hixson, *Maria*, 20 May 1834, HLHU; Shattuck diary, *Frances*, 14 February, 1851.

[50] Shattuck diary, 5 January 1852. Shattuck acted out the role of a man Charles Rosenberg has called the Christian gentleman. This individual, prominent in bourgeois circles, was an "athlete of continence, not coitus." See Charles E. Rosenberg, "Sexuality, Class and Role in 19th Century America," in Elizabeth H. Pleck and Joseph H. Pleck, eds., *The American Man* (Englewood Cliffs: Prentice-Hall, 1980), pp. 229–31.

physiognomies were deemed most beautiful and, for these abstemious seamen, most diabolically tempting. Polynesian women seemed to fall into this category most regularly. John Cleland, seaman aboard the *Ceres* in 1836, asserted that Tahitians were the "best looking kanakas." The reasons, he said, were clear. They had "regular" features, their lips were "thin," and their teeth were "even." Richard Hixson, who described Maui prostitution as a "disgusting business," was nevertheless moved to allow himself to be physically close to these natives: "I went in to many of there huts, no sooner in, than the females begin to examine every part of my person, express a great deal of curiousity – They are very particular in there examination, sometimes they go, or attempt to, so far that they infringe on my modesty." Hixson made no attempt to thwart such investigation, however.[51]

William Allen said that he and similarly "moral" whalemen were in continual danger in the Marquesas, and his breathless description of Marquesan women explains why: "Many of them are really handsome with fine glossy long black hair curling over their shoulders and decorated with a profusion of beautiful sweet scented flowers beautiful jet black eyes glowing with passion a beautiful set of teeth and clear skin much lighter than a mulattoes." In contrast, Allen quickly distanced himself from the women of Callao (Peru), revealing a virulent double prejudice. "The next thing I saw," he said, in touring this city, "was some – let me see – in our country they call animals that wear bonnets and long togs, women, yes, I saw women but such women!! they were as *black* as my hat or blacker, and about as big round as they were long, and I see one fall down and instead of stopping as common folks do, after she was down she ended over and came on to her feet just like a cask."[52]

A sense of race and sex superiority also found a comfortable place in the mind of George Blanchard. This whaleman acknowledged the appeal of St. Jago natives to other seamen but dismissed with repugnance the idea that he himself would ever make love to these women: "Love could never nestle on the thick Black Lips of a Portugee niggar.... Saving their faces (the best resemblance to which is their imitative companions of the woods, the monkeys) the young ladies ... might rival the finest figures in our own Country. In purchasing one of these Animals, you don't buy a Pig in a Poke, you see your bargain." Blanchard linked his distaste for St. Jago with his perspective on women in general. The island, he felt, was "wretched, naked, disgusting." It was, he said, "like a womans mind arrived at its utmost limits early in its existence – it never has improved – and like a woman – never will, untill it changes masters."[53]

Although sailors like George Blanchard avoided intimate contact with native

[51] Diary of John Cleland, *Ceres*, 31 March 1836, MM; Hixson diary, 20 May 1834.
[52] Allen diary, 9 November 1843; 4 April 1842.
[53] Blanchard diary, *Pantheon*, 26 December 1842.

women, they did not indiscriminately condemn indigenous peoples. Those regions that had felt the "ameliorating" influence of white missionaries, for instance, were felt to have potential for improvement. "The Canainkas," wrote Elias Trotter in 1846, "are fast rising in civilization & the artes . . . they ape wonderfully the white and being as they are now thrown in contact with the cunning Yankee, they learn much of his tact & manner." And these seamen took a certain morbid interest in "savage" ways. Not so disgusted by native practices like cannibalism that they could not be good tourists, they bargained enthusiastically for war clubs, spears, bows, and arrows. And they took a detailed interest in those naked bodies that made them so uncomfortable. Marquesan tattoos received particular attention in whaling journals, as did the appearance and positioning of Polynesian loincloths in general. William Wilson, who visited an island in the central Pacific in 1850, took investigative reporting to special extremes when he detailed the menstruation of women there and commented on the nature of their discharge.[54] As some of America's earliest ethnologists and scientific racists, these traveling whalemen diligently recorded and ranked physiologies in order to confirm in their minds Anglo-Saxon supremacy.[55]

In their social surveys, Victorian whalemen took in more than native women. They readily compared their own work ethic with that of men ashore, for instance, and condemned the "indolence" that they witnessed in port. They also denigrated the "unwashed" people they saw around them. Despite their own recent involvement in greasy work, they clung tightly to bourgeois ideals of hygiene, and as they walked the paths and crooked streets of places like Talcahuano, Callao, and Tumbes, they shrank from the sight of the dirt they saw everywhere.[56]

Middle-class whalemen saved their most heartfelt condemnation for those people with triple "deficiencies": dark skin, a seemingly retrogressive work ethic, and Catholicism. This is hardly surprising, given the origins of these sailors in a land swept by a new economic order, Protestant revivals, and intense racial conflict. In fact, the comments whalemen made about peoples in Latin America, the Pacific, and elsewhere echoed the disparagement that war-

---

[54] Diary of Elias Trotter, *Illinois*, 20 March 1846, Log 1005 A & B, WMLOD; Wilson diary, 19 January 1850.

[55] George M. Fredrickson, *The Black Image in the White Mind: The Debate on Afro-American Character and Destiny, 1817–1914* (New York: Harper & Row, 1971), pp. 71–96, provides a good overview of the rise of scientific racism in the antebellum period.

[56] On the intensifying work ethic among the "Yankee bourgeoisie" at this time, see Daniel T. Rodgers, *The Work Ethic in Industrial America 1850–1920* (Chicago: University of Chicago Press, 1978), p. xiii; on the cleanliness campaign in American history and its relation to whalemen, see Chapter 3 of this book. For examples of sailors who censured Atlantic and Pacific populations for their alleged indolence and dirtiness, see the Diary of James Allen, *Alfred Gibbs*, 21 March 1871, PPL; Cleland diary, 10 July 1834; Hixson diary, 9 March 1833; Diary of Stephen Curtis, *Mercury*, c. October 1841, NBFPL.

mongering Americans directed at the men and women of their neighboring republic, Mexico, and which helped them rationalize military conflict.[57]

Whalemen denounced in particular the volatility of the people of Peru and Chile. When William Allen walked out to a battlefield near Lima in 1842, he saw evidence of unbridled passion everywhere. He saw skeletons of "at least one thousand men laying round, and remnants of their dress and parts of their armour, old rusty bayonets [and] broken swords." They were, he said, the Peruvian remains of men who had been fired on by their own compatriots. This "bloody cannibalism" and fighting were typical, he felt, of this "degraded" population. "This is a specimen of Peru," he said, "one of the richest countries in the world." He noted that even as he wrote Peruvians were fighting "between themselves" in the interior. Allen, spokesperson of that society that would slaughter itself in a Civil War two decades later, was horrified.[58]

Sailors whose sensibilities were discomforted by dirt, "wasteful" behavior, and Catholic "bigotry" were relieved to leave port. Richard Hixson was "not sorry" to weigh anchor in Callao, for "our deck has been covered day & night by Spaniards and a Spaniard I do not like." James Allen hoped to sail from Talcahuano, "this land and water of fleas" as soon as he could. James Haviland was finally out of sight of Hawaii and wrote "I can assure you I am heartily glad of it." Unlike the majority of sailors for whom port stops had meant freedom and pleasure, and who had found some commonality with native inhabitants, this minority had endured what seemed like unending social assaults, and they weighed anchor with little reluctance.[59]

Given the presence of sensitive Victorians among whaleships' crews, as well as the presence of men who shared at least some of their opinions, it is little wonder that whalemen did not make a concerted effort to challenge the prohibitions of missionaries. Too many of them agreed, wholeheartedly or in part, with what these reformers were all about. The deepwater brotherhood, which had stood against authority at sea, frequently gave way in port to dissension. Cultural and class divisions, which had diminished on shipboard, rose anew ashore. And yet political fraternity was not an utter stranger to sailors on liberty. On at least one occasion, whalemen did stand together to challenge authority. Significantly, this involved a protest not against advocates of social reform but against the police establishment. And it involved not primarily clashes of morality but issues of life and death. In many ways, therefore, it mirrored contests at sea.

---

[57] On American attitudes toward the peoples of Mexico, see Reginald Horsman, *Race and Manifest Destiny: The Origins of American Racial Anglo-Saxonism* (Cambridge, MA: Harvard University Press, 1981).   [58] William Allen diary, 4 April 1842.

[59] Hixson diary, 13 March 1833; James Allen diary, 1 April 1871; Haviland diary, 15 November 1857.

In early November 1852, a seaman from the ship *Emerald* was clubbed to death, allegedly by a Honolulu constable. Sailors, many of them whalemen, joined together in bands of 400 to 500 to punish the offender. When authorities refused to deliver the constable over to the seamen as they demanded, the mariners took matters into their own hands. They burned the police station and the harbormaster's house, and attacked several closed saloons. They also rescued a sailor who had been previously arrested. When the governor proclaimed martial law, locked up rioters, and imposed a curfew, order was restored. At the same time, the sailors were assured publicly that the constable would be tried for the crime.[60]

Mariners, then, occasionally combined on shore as a rebellious force. So, too, did they sometimes try to keep together their social community. Foremast hands went their separate ways on liberty, but they continued to remind one another that they remained a ship's crew. If a man left his shipmates to conduct private business in port or even if he abandoned his sailor garb for some shore togs, he was subject to the same sorts of rebuke that he might have encountered in the forecastle itself, particularly when he returned aboard. He learned that a sailor simply did not choose a visit to a bethel or to a private reading room as the only stop in a foreign port. If a man wanted to keep peace with his crew, he also visited a shore side bar, where he and his shipmates would signal the ongoing strength of their camaraderie with the shared purchase of rounds of rum.[61]

---

[60] W. D. Alexander, *A Brief History of the Hawaiian People* (New York, 1891), p. 274; "The Whalemen's Shipping List and Merchants' Transcript," 18 January 1853.

[61] On this sort of social pressure in port, see the Haviland diary, 18 November 1857. Richard Henry Dana, *Two Years Before the Mast* (1840; reprint: Los Angeles: Ward Ritchie Press, 1964, ed., John Haskell Kemble), p. 118, also describes the difficulties of a middle-class sailor on shore with his shipmates.

7

Sailors, Sweethearts, and Wives

Gender and Sex
in the Deepwater Workplace

It is no place for a woman on board of a whaleship.

James Haviland, *Baltic*, 1856[1]

When men went whaling, they entered a blue-water brotherhood. King Neptune initiated them into a distinctive working fraternity, and the demands of labor and life at sea, particularly in the forecastle, strengthened men's allegiance to one another. Whaling represented brotherhood in a more basic way. These sailors were exclusively male. A few women may have sailed on whaleships in disguise, but a survey of crew lists, ship logs, sailor diaries, and secondary literature indicates that the female working sailor of whaling's golden age was essentially a contradiction in terms.[2]

The single-sex nature of the whaleship invites wide-ranging speculation on sailors' alleged difference from mainland men. This chapter, and to some extent the next, engages in such broad inquiry. It considers whether or not whalemen sought seafaring as a special refuge from women and domesticity and if they stood opposed to prevailing ideals of marriage and the family. It examines the effect of long voyaging on social relations between men and women and on male self-esteem. It considers the implications for men of breaching the traditional

---

[1] Diary of James Haviland, *Baltic*, 3 November 1856, PPL.

[2] Sex segregation on American deepwater vessels may have been a diminishing phenomenon. Judith Fingard has described the employment of women as stewardesses aboard Atlantic Canadian vessels in the 1880s. See *Jack in Port: Sailortowns of Eastern Canada* (Toronto: University of Toronto Press, 1982), p. 57. Melinda Campbell has noted women among Great Lakes crews in the mid- to late nineteenth century: "The Backwoods Seaman: Great Lakes Merchant Sailors in Green Bay, Wisconsin, 1847–1880" (paper delivered at the Jack Tar in History Conference, Halifax, Nova Scotia, 27 October 1990).

gender division of labor (men at sea did "women's" work). Finally, it briefly explores the elusive subject of sexual behavior and romance between men on shipboard and considers whether or not whalemen might have been distinctly tolerant of homoerotic contact.

I

There were few working women on American whaleships, but that does not mean that women did not sail in whalers. Captains' wives went to sea increasingly in the nineteenth century. As tourists, onlookers, "mothers" to the ship's crew but most important as true wives, these daring women moved themselves and sometimes their children into ships' aftercabins.[3] Their presence provoked some distinctive commentary from lesser officers and foremast hands. George Coffin, sailing before the mast on the ship *Niger*, wrote to his sister that his captain's wife, who had a nineteen-month-old child aboard, "is liked very much by all." Other shipmasters' spouses, whose associations with the crew ranged from dispensing gifts to softening ship discipline, elicited similar praise, particularly from cabin boys.[4] Many of the men who discussed sailing wives, however, were vehemently hostile toward bluewater women. "I think you wished me very ill when you told me you were glad that the Capt was going to carry his wife," Marshall Keith commented, "for she is a source of trouble." Robert Weir echoed his remarks. Mrs. Macomber of the *Mary Ann* was, he said, a "disgusting woman."[5]

Two other sailors vented their contempt for masters' wives with remarkable specificity and focused their indignation on the women's noses. A Mrs. Brown

[3] *The Whaleman's Shipping List and Merchants' Transcript* in 1853 estimated that over 15 percent of whaling masters sailed with their wives; see "Lady Whalers," 1 February 1853. See also Lisa Norling, "'I Have Ever Felt Homeless': Mariners' Wives and the Ideology of Domesticity 1780–1880" (paper presented to the Philadelphia Center for Early American Studies, 30 October 1987), and Joan Druett, "More Decency and Order: Women and Whalemen in the Pacific," *Log of Mystic Seaport* 39 (1987): 67. For an extraordinarily thorough examination of the experience of whaling wives, see Druett, ed., *"She Was a Sister Sailor": The Whaling Journals of Mary Brewster 1845–1851* (Mystic, CT: Mystic Seaport Museum, 1992). The presence of captains' wives may be tied to a number of developments: the increasing length of whaling voyages in the 1800s; reformers' interest in domesticating seafaring; the rising importance of companionship in marriage. On the last subject see Carl N. Degler, *At Odds: Women and the Family in America from the Revolution to the Present* (New York: Oxford University Press, 1980).
[4] George F. Coffin to Phebe Ann Hanaford, 16 October 1859, NHA; Diary of Mary Stickney, *Cicero*, 24 December 1850; 17 January 1851, PPL; Diary of John States, *Nantasket*, 18 November 1845, MSM; Diary of John Spooner, *Mt. Wollaston*, 27 April 1861, NBFPL; Diary of William Abbe, *Atkins Adams*, 29 July 1859, Log 485, WMLOD.
[5] Diary of Marshall Keith, *Brewster*, c. 1864, PPL; Diary of Robert Weir, *Clara Bell*, 26 January 1856, MSM.

of the *Northern Light* was, according to seaman Lewis Williams, blessed with a
striking facial protuberance. "My descriptive powers fail me here and I will say
it was a Nose but what a Nose. I shall not proceed any farther with my
description for that Nose looms up before me in such magnificent proportions
I can see nothing else." Edward Mitchell was not quite so dumbstruck when it
came to describing the nose of Elizabeth Harriman, his shipmaster's wife. He
even put his contempt into verse: "She walks the deck with majestic grace.
With her cherry picker nose all over her face."[6]

Historian Eric Sager recently asked whether seafaring was, among other
things, an escape from the company of women.[7] Certainly the foregoing sailor
invective suggests that some men did not welcome the presence of women on
deepwater. But other whalemen suggest that the matter was more complicated.
They tell us that *escape* may indeed be a relevant term but that it depends on
what women – masters' wives, mothers, sweethearts, or sexual partners – and
what men – married or unmarried seamen, foremast hands or shipmasters,
white seamen or men of color – we are talking about.

Sailor hostility toward shipmasters' wives may have been fueled, for instance,
by a complicated intertwining of class, gender, and age antagonism. Many
aftercabin women endangered the tenuous solidarity of foremast hands with
their plans to reform sailors. At the same time that veteran seamen struggled to
democratize and unify the forecastle and to substitute a seafaring family for a
shore- or female-centered one, shipmasters' wives interceded and lured men's
allegiances aft.

These women actually accomplished this in several ways. Not infrequently,
they identified forecastle sailors whom they deemed "respectable" and rewarded
them with presents. A Mrs. Jernegan gave Orson Shattuck, a man of "refine-
ment" in her husband's crew, some "poultices and oyster soup" when he was
stricken with quinsy but did not honor the rest of the men with similar attention.
Mrs. Wilson aboard the *Atkins Adams* singled out Harvard student William Abbe
for various special favors and sent him pieces of cake, a bottle of mustard, and
a pot of grape preserves, and even lined a new hat for him. Abbe himself
admitted that Mrs. Wilson demonstrated a "partiality for me above the other
hands." Such demonstrations of favoritism helped to make enemies for William
Abbe among the ship's crew, and they no doubt did not win Mrs. Wilson any
special popularity.[8]

---

[6] Diary of Lewis Williams, *Gratitude*, 13 January 1860, PMS; Diary of Edward Mitchell,
    Ivanhoe, 16 April 1867, PMM.
[7] Eric Sager, *Seafaring Labour: The Merchant Marine of Atlantic Canada 1820–1914* (Mon-
    treal: McGill–Queen's University Press, 1989), p. 237.
[8] Diary of Orson Shattuck (under the pseudonym Charles F. Perkins), *Eliza Mason*, 8 August
    1854, Log 995, WMLOD; Abbe diary, 18 November 1858; 5 December 1858; 2 January
    1859.

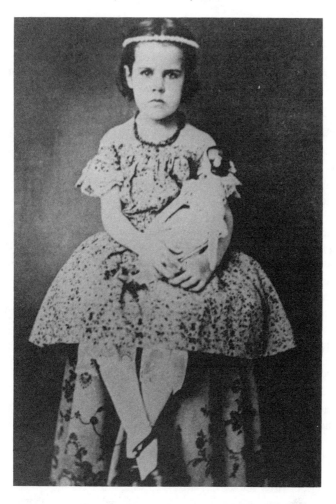

Minnie Lawrence accompanied her parents, Mary Chipman Lawrence and
Captain Samuel Lawrence, aboard the whaleship *Addison* in the late 1850s.
An ambrotype was taken of Minnie in the Hawaiian islands on one of the
many stops the *Addison* made between whaling cruises.

Women who sailed did not always select individual sailors in their attempts
at social reform. They sometimes made broad efforts to better the habits of
seamen. Mary Chipman Lawrence of the *Addison* was among several captain's
wives who brought boxes full of Bibles and New Testaments aboard ship and
distributed them generally to the men before the mast. Lawrence herself employed her daughter Minnie as her missionary, filling Minnie's own carriage

with Bibles and sending her wheeling off toward the forecastle. She was heartened
to see the carriage come back empty. "It may be," she exulted, "we can do some
good through her."[9]

The visit of Minnie Lawrence to the forecastle suggests one reason women
on shipboard might have provoked antagonism among some seamen: women
undermined opposition and collectivity before the mast and drew some sailors
willingly into their cultural milieu. It suggests another reason as well. Women
(and children) invaded sailors' space at will, thereby emphasizing sailors' (including junior officers') subordinate rank. Women who sailed with infants and
toddlers aboard, as many did, did not maintain quiet profiles aboard ship, and
they enlarged the presence – both material and political – of the master.

Of the dozen or so sailors who felt frustration on this score, none can better
the angry vehemence of Abram Briggs, a fourth mate aboard the *Eliza Adams*.
According to Briggs, the *Eliza Adams* barely had a chance to cruise for whales
in 1873 because needs of the family aboard took priority over the hunt. In 1873
alone, Briggs noted, the whaler stopped ashore in Australia four times in six-
and-a-half months to attend to the pregnant Mrs. Hamblin and then her
newborn.[10]

To Abram Briggs, the captain's wife was nothing other than a "cow," and
when she and her "calves" came aboard the ship, they turned the vessel into
"hell afloat." A number of things particularly disturbed Briggs. The captain and
his wife stayed in bed late and kept the officers waiting for their breakfast. The
captain's wife told the man at the wheel not to strike the ship's bell too hard:
it woke the "calf." The master's son rode his wagon over the heads of the watch
below, and the children ran on deck in the daytime. When a sailor complained
about the disturbance, Briggs reported that the captain's wife "told them to
make as much noise as they like, for father owned the ship." And, finally, the
captain expanded his privileged station by rebuilding the cabin and the house
on deck to accommodate his family. When, on the other hand, the boatsteerers
asked the shipmaster for some boards with which to make the cabin boy a bunk
(he was sleeping on a sea chest), the captain allegedly asserted that lumber was
"getting scarce" and refused to supply some. Abram Briggs hoped that he
would see the happy day when Mrs. Hamblin and her children were "in the
heart of Cape Cod burried up to their necks In sand."[11]

Abram Briggs may have had a lower threshold of anger than other mariners,
but the circumstances that helped him cross it were not uncommon. A number

---

[9] Diary of Mary Chipman Lawrence, 2 April 1857, in Stanton Garner, ed., *The Captain's Best
Mate* (Providence: Brown University Press, 1966), p. 22.

[10] Diary of Abram Briggs, *Eliza Adams*, 27 September 1873, Log 940, WMLOD.

[11] Ibid., 30 April 1875; 12, 27 July 1874; 4 February 1873; 29 November 1872; 12 November
1873.

of children shared shipboard space with whalemen and they participated in a wide range of activities. They sat in the middle of sailmaking, they climbed the ship's rigging and fell overboard, and they "helped" to cut in whales. They also kept pets (in one case more than thirty birds), which occasionally went into the sea, and which were rescued only by attentive and selfless acts of seamen.[12]

Shipmasters' wives and children exacerbated the inferior rank of sailors by inconveniencing them. They also irritated and embarrassed seamen by observing them. "I want to see everything that is going on," explained Mary Chipman Lawrence. What Mary Lawrence meant is not clear, but according to a few sailors, "everything" meant everything. "The carpenter," noted Abram Briggs, "has put a long window In the forward part of the house, so Mrs. Hamblin can set down & look whats going on on deck, who goes over the bows, or to the Urine barrell." Sailors were thus disadvantaged by a woman who could oversee their intimate acts while her privacy was relatively protected. They may have been further annoyed by women who apparently had no productive role at sea (the steward and the cook did the housekeeping and the cooking) and who seemed to be merely onlookers. The mate of the *Nautilus* summed up various grievances when he swore at the master's wife as she watched the men at work. "Damn you," he declared, "look if you want to."[13]

Mariners troubled by the presence of a captain's spouse alluded to a further problem. A shipmaster's wife, with her ability to fragment the forecastle, impede whaling work, and intrude on male privacy, seemed to draw power away from – emasculate might not be too strong a word in this case – the master. Although shipboard politics were fraught with conflict, all sailors to some degree acknowledged the master's power and committed themselves to it. In return for their allegiance, they expected the captain's focus and energies to be directed to them and their work. A wife aboard was a complicating distraction. And whaling women not only committed this specific sin but transgressed on a wider social level as well. They subverted the patriarchal order. Mrs. Elizabeth Harriman, the woman who allegedly walked the deck "with her cherry picker nose all over her face," was the "Captain of the ship *Ivanhoe*," a ship that had a "double skipper." Mrs. Hamblin, wife of the master of the *Eliza Adams*, stayed in bed late, kept officers waiting for their breakfast, and was prone to childbearing. She was someone, according to the irrepressible Abram Briggs, who had taken complete charge of her husband. "When I sail with another woman," wrote Briggs, "it will be after I quit going to sea, and especially with one, one that wears the breeches, far from it (God preserve me from that)."[14]

[12] See Margaret S. Creighton, "The Captain's Children: Life in the Adult World of Whaling, 1852–1907," *American Neptune* 38 (July 1978): 203–16.

[13] Lawrence diary, 14 March 1857, in Garner, *Captain's Best Mate*; Briggs diary, 29 November 1872; Diary of Lucy Smith, *Nautilus*, 7 June 1870, PPL.

[14] Mitchell diary, 16 April 1867; Briggs diary, 23 May 1875.

The most successful sailing wives from sailors' perspectives, then, were those who kept to themselves. One of the few charitable comments to emerge from the diaries on this subject was one noted by whaleman Lewis Williams about his sister-in-law, who had gone to sea with her husband. Williams met a man at a gam who had sailed with Eliza Williams and spoke highly of her, giving her "an excellent name." What was this woman's key to success at sea? "Always minding her own business," asserted her brother-in-law, with conviction.[15]

The negative feelings that many sailors expressed about shipmasters' spouses were tied to the fact that these women represented both the power of the shipmaster and the (threatened) power of women. These sentiments were also related to the fact of sailors' youthfulness and to their use of the sailing voyage as a rite of passage into manhood.

Many young whalemen looked forward to what the voyage promised them in terms of economic independence and providership. They strove to enhance toughness and daring – two important masculine credentials – during the whale hunt and used their time off duty to develop the "manly" habits of smoking, swearing, and gambling, and (on shore) drinking and having sex with strange women. Victorian sailors, as we have seen, looked to life on a sailing ship as an enforced lesson in self-control.[16]

Central to making "men" out of boys was discarding feminine attributes. Veteran seamen played a significant part in steering novices in the forecastle away from womanly behavior. They specifically sought to eradicate traces of delicacy, sentimentality, and tenderness, qualities associated with the middle-class female ideal.[17]

Being a "man" in America around 1850, at sea and ashore, with a few exceptions, meant the ability to sublimate sentimentality and emotion. As John Faragher contended in his study of nineteenth-century sex roles, a male who cried was seriously impaired. In Faragher's words, "the male horror of tears was

---

[15] Diary of Lewis Williams, *Othello*, 22 May 1860, PMS.
[16] Working-class and middle-class masculine ideals are discussed in Elliott J. Gorn, *The Manly Art: Bare-Knuckle Prize Fighting in America* (Ithaca: Cornell University Press), pp. 140–42; Mary P. Ryan, *Cradle of the Middle Class: The Family in Oneida County New York, 1790–1865* (Cambridge, UK: Cambridge University Press, 1981), pp. 145–85; E. Anthony Rotundo, *American Manhood: Transformations in Masculinity from the Revolution to the Modern Era* (New York: Basic Books, 1993). The acquisition of minor vices among young men is examined in John C. Burnham, *Bad Habits: Drinking, Smoking, Taking Drugs, Gambling, Sexual Misbehavior, and Swearing in American History* (New York: New York University Press, 1993).
[17] Ryan, *Cradle*, pp. 188–9; Ruth Bloch, "American Feminine Ideals in Transition: The Rise of the Moral Mother, 1785–1815," *Feminist Studies* 4 (1978): 116. The classic description of the ideal Victorian woman is found in Barbara Welter, "The Cult of True Womanhood: 1820–1860," *American Quarterly* 18 (1966): 151–74.

well founded, for cultural consensus about masculine strength and feminine weakness meant that a public display of sentiment might mark a man indelibly."[18]

Young men took plenty of lessons in stoicism within the forecastle. They learned from experienced sailors that it behooved them to adopt a veneer of steely fortitude. When New Hampshire native Orson Shattuck became watery-eyed over the thought of home and his sisters, he knew that if he did not get a grip on himself, he would be in serious trouble with his shipmates. "I must stop this," he insisted to himself, "for I can hardly hide the tears from the rough men around me." When toughness eluded a sailor named Smith aboard the *Sunbeam* in 1868, he suffered for it. "His love for his Grandmother is very strong," explained shipmate Silliman Ives. "A few days since he was discovered sitting upon his chest bathed in tears, weeping excessively over a bed quilt which he informed his shipmates was the gift of his Grandmother, who was upwards of ninety years of age and in all probability he would never see her again. A strong proof of filial affection, a commodity not very troublesome to sailors as a general thing." For his emotional display, Smith was subject to his shipmates' jeers and labeled the "ship's fool."[19]

Neophyte sailors were thus taught to suppress their tears in public. Men who successfully learned manly bravado made a show of their distaste for sentiment and the "soft" behaviors of women. Nathaniel Morgan and his shipmates, for example, shared utter contempt for shipmasters' wives who met each other at gams and who showered each other with kind attentions:

> Capt S and wife, and Mr King and wife already to come aboard – they came aboard and stayed all day – and such a time – more affectation and soft soap than would fill a slush table Nobody seemed to live on this globe but they – and no dinner for us until 2 oclock – They left at 7 pm amid such a scene of soft soap and kissing it makes us all sick – Agreed to keep company and have another gam tommorrow – go ahead.[20]

By making their contempt for affection public, sailors signaled to one another not only their indifference, but also their mature masculinity. In private, though, seamen gave way to the feelings that were still there. Sometimes they cried,

---

[18] John Mack Faragher, *Women and Men on the Overland Trail* (New Haven: Yale University Press, 1979), p. 92. Other analyses of masculinity have shown that for a few social groups, expressions of intense affection were normative. See E. Anthony Rotundo, "Romantic Friendship: Male Intimacy and Middle Class Youth in the Northern United States, 1800–1900," *Journal of Social History* 23 (Fall 1989): 1–25; and Donald Yacovone, "Abolitionists and the 'Language of Fraternal Love,'" in Mark C. Carnes and Clyde Griffen, eds., *Meanings for Manhood: Constructions of Masculinity in Victorian America* (Chicago: University of Chicago Press, 1990), pp. 85–95.

[19] Shattuck diary, *Frances*, 19 October 1850; Diary of Silliman Ives (under the pseudonym Murphy McGuire), *Sunbeam*, 10 July 1868, Log 618, WMLOD.

[20] Diary of Nathaniel S. Morgan, *Hannibal*, 22 November 1850, MSM.

although they did so quietly, and the salt that fell from their eyes even then was not a sentimental display, but new, "manly" tears. (Presumably these were firmer, more solitary drops.) Under the cover of darkness, too, hardened seamen admitted that they needed care and affection and that they yearned for the nurturing touch of mothers and sisters. Mariners afflicted with sickness particularly ached for the healing presence of a concerned female. "The first thing that rushes into the recollections of a Sailor that is smitten with disease at sea is his Mother," noted one whaleman in 1864. "She clings to his . . . affections in the Midst of all the forgetfulness and hardihood induced by a roving life."[21]

The journals of married seamen, usually officers and black men, also reveal that sailors applied the lessons of the forecastle mostly to public life. Several men cried to themselves behind closed cabin doors, confessing their sentiment to their journals or in letters to women at home. Captain Edmund Jennings grew nostalgic for home on a Thanksgiving day in 1877, and "went into my room and had a good crying spell." Marshall Keith, mate aboard the *Cape Horn Pigeon*, felt like crying one day when he saw a "ship bound home." "Hard hearted sailors," though, "would laugh at me." But later, in the privacy of his cabin, he "cryed all the afternoon" after a disagreement with his shipmaster.[22]

Whalemen sought to indoctrinate novices in public hardiness and to strip them of their "female" sides, but it was clearly something of a struggle. When we consider some of the reasons seamen expressed hostility toward women who sailed, one might be that these women exhibited behavior that mariners formally proscribed but secretly still practiced. Their presence may have reminded whalemen of the distance they had yet to travel on the way to true toughness.[23]

II

Young whalemen may not have been hospitable when it came to shipmasters' spouses, but they were distinctly generous when it came to sweethearts ashore. Most seamen made it clear, in fact, that their commitment to tough masculinity on the one hand and to fraternity on the other did not mean a severing of male–female ties. On the contrary, they suggest that the enhancement of manliness had as its object a future female audience, and many of them longed for the day when the ship would set them on shore with the means of "getting spliced." For these men, seafaring represented no real rejection of women or of their

---

[21] Diary of Marshall Keith, *Brewster*, c.1864, PPL. Men who cried in a distinctly "manly" way included George Blanchard, *Solomon Saltus*, 6 August 1845, PC–TB.

[22] Diary of Edmund Jennings, *Mary and Susan*, 29 November 1877, PPL; Diary of Marshall Keith, *Cape Horn Pigeon*, 2 August 1866; 18 October 1866, Log 371, WMLOD.

[23] Mark Carnes discusses the rigidity of gender roles in nineteenth-century America and the ways and means men found for expressing a dual female–male nature. See *Secret Ritual and Manhood in Victorian America* (New Haven: Yale University Press, 1989).

society but, rather, a temporary retreat: a period when a man gathered his forces, shaped a distinctly male demeanor, and secured the means of a livelihood.

But the process of becoming a man at sea, far away from women ashore, was fraught with special difficulties. Sailors who were tied emotionally to sweethearts at home sometimes alienated their shipmates. And whalemen who went to sea to help secure a solid place in the patriarchy found that voyaging put women out of sight and out of control. Mariners used their creative imaginations to resolve the social contradictions of their peculiar profession, but with mixed results.

Many whaling hands left the shores of home at the height of their courting years, and as they spoke of their ardor for hometown sweethearts or "meeting gals" or of their fiancées, their journals fairly steamed with romantic heat. They recorded sweet memories of flirtations and trysts and wrote rhapsodically of sleigh rides and spring rambles through the greening countryside. They recounted the pleasures of maybaskets and valentines, dancing parties and church gatherings.

These whalemen left an America where romantic love was on the rise. Tied in with expanding economic and religious individualism was a growing opportunity for young men and women to choose companions and mates by listening to their hearts. For whalemen, however, these developments meant new challenges, for closer relationships between couples meant more suffering as whaling voyages grew longer and longer.[24]

Whalemen at sea in the mid-1800s did their best to keep romantic links alive. They pinned their brightest hopes on the saltwater mail system. They knew that midocean mail delivery was random at best and that outbound letters, often addressed with nothing more specific than "The Pacific Ocean," could take long years to reach their mark, if they arrived at all. Yet they counted on love letters to arrive by passing ships or in foreign ports, and then became seriously discouraged when they did not hear from sweethearts at least once in a while. Sailors aboard the *Smyrna* were "hurt and disappointed" not to get letters from a vessel that had sailed recently from home. William Taylor, for one, went into a dudgeon when he did not hear from "Mary Ann." "She has forgoten me," he lamented, "or perhaps is better employed than writing to one she thinks so little about, all the rest have letters two or three from their friends and relations, I think it is very hard." Taylor was so mystified by his lack of mail that he actually rowed over to the mail-bearing ship and personally "overhauled her

---

[24] On the rise of romantic love, see Ellen Rothman, *Hands and Hearts: A History of Courtship in America* (New York: Basic Books, 1984), p. 31. Lisa Norling discusses the parallel developments of romantic love and lengthy voyaging in "The View From Shore: Women's Perspectives on Whaling" (Paper delivered to the Old Dartmouth Historical Society, 18 April 1989).

letter bag." He was left, however, to "chew the cud of disappointment." Later he announced that his lack of letters could only mean that Mary Ann had forgotten him and that "every tie of love or friendship for me was broken" and that he might as well "die as live." Taylor had to eat his words, however, for he received two letters from the accused party the following day, and he had to "take it back what I said last night."[25]

Benjamin Boodry, the *Arnolda*'s mate, underscored the importance of letters with the claim that he had "read [his] letters over again for the 7000th time have got the whole 25 of them by hart." Few sailors were as lucky as Boodry, however, and sailors easily felt forgotten. At least two unhappy seamen refused to give up the comforts of mail, though, and they endeavored to create the trappings of a lover's correspondence. Walter Brooks of the *Gladiator* hoped to buy a letter from his shipmates. And on the *Ann Parry* in 1847, Ezra Goodnough found a shipmate willing to pay a heavy price for one of Ezra's hometown letters: "I sold a letter that I received from a young lady of Salem and the only one to that I have received this voyage for two heads of tobacco, it being a very scarce article."[26]

Seamen who had close, amorous ties to shore looked to letters as the best means to keep those connections strong, but they also relied on gamming gossip and homebound ships to report back their concerns and interests. Others tried to keep up their end of the romantic discourse with their journals. John Wady, a Fall River native, was "Home Sick, Sea Sick, Love Sick, Heart Sick." He noted little of his seafaring experiences in his diaries but kept a careful account of his dreams of Hope Ann Sanford. Marshall Keith had an active social life aboard the *Brewster*, particularly in his sleep, and he kept his dreams alive by recording them. On one occasion, Keith's subconscious sent him to one Helen Purrington's house where he stood watch on deck. It must have been tiring work, for Helen appeared and solicitously offered to support him. "I guess I can hold you," she explained. Keith accepted, naturally, and, taking even further advantage of her kindness, was able to "put my hand on her Applicobation." As luck would have it, "she gave Right in I felt her all over and was having a High time." But who should appear into these wondrous fantasies but consciousness itself, along with the fourth mate, who woke Keith to his sad reality. "If It had not been for him," lamented Keith, "the Lord only knows what I would have done to Helen."[27]

[25] Diary of Charles Chase, *Smyrna*, 1 January 1857, Log 655, WMLOD; Diary of William Taylor, *South Carolina*, 15, 16 July 1836; 2, 3 September 1836, KWM.

[26] Diary of Benjamin Boodry, *Arnolda*, 1 May 1853, Log 619A, WMLOD; Diary of Walter Brooks, *Gladiator*, 26 February 1854, MSM; Diary of Ezra Goodnough, *Ann Parry*, 23 March 1847, PMS.

[27] Diary of John. H. Wady, *Clifford Wayne*, 25 March 1841; 29 December 1842, PPL; Diary of Marshall Keith, *Brewster*, 25 January 1864, PPL.

Romantic reveries, although offering some comfort to lonely seamen, were nevertheless problematic. Sailors who mourned over lovers seemed to threaten the success of a voyage with an early desertion. Two years after Marshall Keith was interrupted in his dreams about Helen Purrington, he pined after Sarah Pope, but his shipmaster chastised him for his fondness. "He thought you was in my mind Most of the time," Keith wrote to Pope, "and that I wanted to see you so much that I was making myself sick to get home."[28]

Sailors who remained heartbound to home, as did Keith, also faced the perennial tensions between allegiances to shore and the demands of the ship's company. Their dilemmas were solved to some extent by shipboard rituals that allowed seamen some limited landward focus. The first of these took place in the forecastle on Sundays. During this ritual, known commonly as the "sailor's pleasure," seamen literally opened their ocean-bound lives to the materials and memories of home. They accomplished this by organizing or "overhauling" their sea chests and by "trimming" their ditty boxes. The latter contained domestic utensils – needles, buttons, thimbles, scissors, and perhaps a New Testament. The sea chest held a number of items that connoted home. It contained "long togs" or home clothes, daguerreotypes, and old letters. After indulging in the material reminders of shore, perhaps for an hour or so, a sailor closed the lid of his chest for the week.[29]

An aftercabin ritual was similarly timed so that it did not seriously impinge either on the work routine or on ship solidarity. On Saturday nights, mariners sometimes set aside time (and alcohol, too, on nontemperance ships) for a toast to "sweethearts and wives." John Alden, a passenger aboard the *Albatross* in 1847, who enjoyed a whiskey punch in toasting the health of sweethearts and "wifes," asserted that he believed it "is an old established custom to devote Saturday evening to this cause, and although it may in some instances be needful to remind one of his duties to those he left behind him – such is not the case with me."[30]

Alden's remarks point out the conflicts that obviously existed between sailors' allegiance to a ship and their attachment to home, and it is echoed in a "Saturday Night" sea song that a whaleman recorded in 1843:

> A sailor loves a gallant ship
> And messmates bold and free
> And even welcomes with delight
> Saturday night at sea.
> One hour each week we'll snatch from care
> As through the world we roam

[28] Keith diary, *Cape Horn Pigeon*, 18 October 1866.
[29] See George Brown Goode, *The Fisheries and Fishery Industries of the United States*, Section V, Vol. II (Washington: Gov't Printing Office, 1887), p. 231.
[30] Diary of John Alden, *Albatross*, 5 June 1847, PMS.

And think of dear ones far away
And all the joys from home.[31]

Restricting homeward thoughts to one or two hours out of 300 amounted to a plan of action by some men. Captain Albert Goodwin, for one, allowed himself to get out his daguerreotypes once a week. For a short time he paraded them before him and he kissed each one. Then he retired them for the next six days.[32]

Shipboard ritual remembered women, but it honored flesh and blood individuals in these carefully circumscribed ways. Generic, objectified women had a more universal and in some ways more legitimate place in deepwater culture. They appeared in sea music, shanteys, artwork, and even in the ship itself. Their presence in shipboard life publicly signified sailors' interest in females, but it also reinforced sailors' commitments to shipmates.[33]

Whalemen tended to depict women as two types: the faithful woman at home and the sexually free female. They glorified loyal women both in the medium and message of their artwork. Enterprising sailors carved whalebone to make scrimshaw gifts for mothers, sisters, and faithful sweethearts. They produced massive numbers of yarn-winders, pie crimpers, hair pins, and work boxes. On these items and on the ubiquitous sperm whale's tooth, seamen carved scenes of tearful partings and joyful reunions that underscored their most fervent prayers. At the same time, they honored the conventional sexual female. They carved toothpicks, whistles, and pipe-tampers in the shape of female legs, and they etched and sketched figures of women half-clothed.[34]

Seamen not only carved scrimshaw to indulge their two fantasies of females, but they also used parts of the sailing ship itself as a vehicle for their social aspirations. Many American vessels in the nineteenth century had female figure-heads. This wooden statue on the prow of the ship could be a reminder of wife, home, and family. According to folklorist Margaret Baker, captains "hated losing a figurehead through storm or accident," and one master reportedly refused to replace his lost figurehead because of its association with domesticity. "Do I seem," exclaimed the offended master, "one who would pick up with another's cast off figurehead? I'd sooner think of taking up with a new wife!"

---

[31] Gale Huntington, *Songs the Whaleman Sang* (Barre, VT.: Barre Publishers, 1964), p. 65.

[32] Diary of Albert Goodwin, *Tuscaloosa*, 28 December 1845, PMS.

[33] Caroline Moseley, "Images of Young Women in Nineteenth-Century Songs of the Sea," *Log of Mystic Seaport* 35 (1984): 132–9 discusses the representation of women in sailors' songs. Gorn, *Manly Art*, p. 142, claims that within nineteenth-century "bachelor subculture" women were depicted as "either pure and virginal or exciting or whorish," and Carroll Smith Rosenberg describes the temptress/virgin images in male purity literature. See "Sex as Symbol in Victorian Purity: An Ethnohistorical Analysis of Jacksonian America," *American Journal of Sociology* 84 (1978): S242.

[34] E. Norman Flayderman, *Scrimshaw and Scrimshanders: Whales and Whalemen* (New Milford, CT: N. Flayderman, 1972), pp. 36–61.

This figurehead of Awashonks, a sachem of the Sakkonet Indians, graced
the prow of a whaleship of the same name, built in 1830. *Awashonks* made
twelve voyages before she was crushed by ice in the Arctic whaling disaster
of 1871. This figurehead had been removed six years earlier.

Baker noted that one figurehead, which sported "high button boots and country
day dress," and which was "sturdy, domestic, redolent of apple pie and warm
kitchens," was a "better reminder of home than a portrait." And another induced a lovesick crew to repeatedly creep forward after dark "to pour their
troubles into [her] sympathetic wooden ears."[35]

[35] Margaret Baker, *Folklore of the Sea* (North Pomfret, VT: David & Charles, 1979), pp.
19–20.

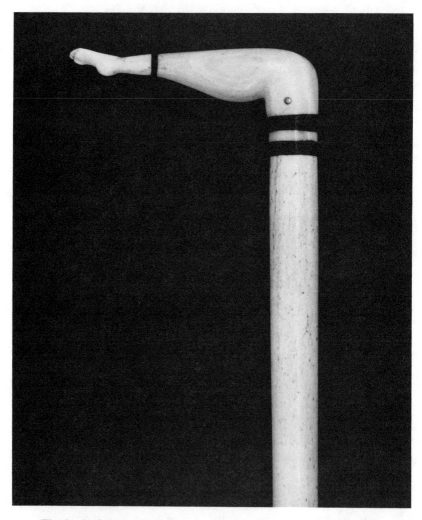

The head of this scrimshaw cane was carved into a shapely and easily graspable female leg.

Owners sometimes used female relatives as models for figureheads and thus sent symbols of domestic, middle-class purity to sea. But, as Margaret Baker again argues, they also often chose female figures who were barebreasted. "It is an ancient nautical belief," she claims, "that a storm will quieten if a woman exposes herself to it *nuda corpore*."[36] Thus some sailors who consulted the

[36] Ibid.

The drawing on this whale's tooth is highly unusual for at least two reasons. Whalemen, judging from their journals, tended to categorize women into pure, pious (marriagable) types on the one hand, and sexually available women, of lower status, on the other. The depiction of a sexually active woman who seems also to be a proper and genteel Christian is distinctive, to say the least. There are also very few pieces of nineteenth-century erotic scrimshaw in museum collections. Was it that Victorian sea captains censored shipboard art? Or that sailors themselves intended most of their scrimshaw for mothers or sisters? Or that pieces like these were not prized (and saved) as family heirlooms?

figurehead under the cover of darkness may have visited a sister or a sexually pure sweetheart, but others might have projected onto this inanimate being their fondest wish for a sexual object.

Sailors' fantasies of women extended to the ship itself. Although no more than a third of the vessels described in this study carried female names, all of them were firmly feminine in gender. Whalemen had some difficulty, though, in projecting sexual agility onto their squat whaling "tubs" and stand in contrast to merchant seamen, who easily sexualized their ships as lively, animate, pretty things. Merchant sailor Isaac Baker compared his vessel under full sail to a "beautiful belle of a ball room." Even in the absence of wind, he insisted, when her sails were listlessly flapping against the ship's mast "still she resembles woman, lovely woman languishing under the effects of ennui and of course 'en dishabille.'" Baker carried the analogy even further. When the wind blew a gale and the ship "scatters the spray around vehemently without respect to persons," it was a "true emblem of the fair sex when hurrying through the labours of a washing day!"[37]

Once again, merchant sailors tended to see in their ship both the woman they could "own" and the woman with whom they might make love. The ship, as Isaac Baker points out, could be a woman enticingly undressed. A sailor could "ride" her and "witness the motions . . . as she plunges & reares her proud head to the sea." He might be "in raptures with her" as she "mounts [the billows] like a thing of life." He could discuss her "bottom" which might be "so full of barnacles that it takes nearly a gale o' wind to set her agoing" or might simply be plump:

> The Taskar is the thing to roll
> O ee roll & go
> Her bottom's round as any bowl!
> O ho roll & go![38]

At the same time the ship could be a sailor's treasured possession. He had a competitive pride in her "good looks," if she had them, and in her sailing abilities. A sailor named Augustus Hamblet made explicit what other mariners expressed more tacitly: "Next to his girl," explained the seaman, "a sailor loves his ship and takes an honest pride in seeing her look neat and trim and never will let any other vessel go ahead of him if he can help it and if they do it puts them (generally) in about as good humour as making a breakfast of vinegar would. But if he beats and goes ahead he is as happy as a clam at high tide."[39] In translating the ship into a girl and the barren ocean into a social main street,

[37] Diary of Isaac Baker, *Taskar*, 29 May 1842, PMS.
[38] Alden diary, 5 June 1847; Diary of John Crimblish, *Palestine*, 31 August 1839, PMS; Baker diary, 11 September 1842; 15 October 1841.
[39] Diary of Augustus Hamblet, *St. Paul*, 29 June 1839, PMS.

these seamen could parade their women past each other, puff themselves up with pride, and compare notes.

Whalemen endowed ships with female qualities, too, but they rhapsodized about them less. Whaling "tubs," which were slow sailers with relatively little canvas spread, and which were round and stubby, did not seem easily suited to sexual analogy. Nor were they as provocative of male pride. Just as whalemen had seemed to fall short when it came to comparison with merchant sailors, these whaling "women" were deemed inferior as well.

Sailors introduced objectified women into ship culture, in reverence or in ridicule, for several reasons. They acknowledged the importance of women on a universal level but protected the seafaring fraternity from the intrusion of particular attachments. They could also gain some measure of symbolic social control. "Woman" in generic form, as the subject of song or art or as the ship itself, could be molded into whatever the mariner wished. The sexy and available woman could validate a man's virility and express at the same time his freedom from social ties at home. ("The sailor is true to his Sal or his Sue," went a capstan shantey, "as long as he's able to keep them in view.")[40] The dutiful, faithful woman demonstrated a mariner's attractiveness, too, but more than that she underscored his ability to keep her in tow. Gale Huntington, who has studied whalemen's songs, claims that one of the most popular lyrics among whalemen was one in which a mariner tests his sweetheart's purity by appearing to her in disguise, attempting to woo her away from her sailor (himself). She remains true to her man. In another popular whaling song, "Susan" on shore demonstrates the ultimate fidelity: she dies of heartbreak when she learns her sailor has been killed:

> The news of his fate with reluctance and sorrow
> The very next day to his Susan they bore
> She heard and frenzy her wits seemed to borrow
> She smiled looked around her and never spoke more.
>
> In the grave with the lad she both lived and died for
> Were laid the remains of the girl he loved dear
> And while to his memory his mates heaved a sigh for
> Each lover will give to his Susan a tear.[41]

The desire to shape a woman to suit a man's fancy was not exclusive to sailors. Yet whalemen felt particularly troubled at their inability to command constancy and its corollary, obedience. One of the most damaging effects of a voyage, they indicate, was that it could be socially disempowering. Although the very act of sailing involved distinct agency, where men did the departing, it seemed to whalemen at sea that once they were abroad on a distant ocean,

---

[40] Frederick Pease Harlow, *Chanteying Aboard American Ships* (Barre, VT: Barre Gazette, 1962), p. 43. [41] Huntington, *Songs the Whaleman Sang*, p. 62.

women at home took the initiative in courting and lovemaking. They were assisted by odious landsmen who had previously lurked offstage, but women themselves, wives included, seemed fully capable of taking social matters into their own hands and adjusting them to their advantage. This included, whalemen were sad to report, sexual affairs. Purity may have been a viable fantasy for mariners, but in reality it was a dim hope. Many sailors felt sure that women at home were not only capable of expressing sexuality but also of switching partners. Captain Samuel Braley remarked upon the degree to which women at home were not unlike the roving sailor: "How many poor Sailors have . . . had unfaithfull wives and they none the wiser for their slips, but it is only tit for tat." "I have been listening," echoed a mate named Marshall Keith, "to hear the captain tell how untrue women were to their husbands while they were away."[42]

Whalemen heard rumors of female infidelity from hometown reports gleaned from shipboard gossip or from letters sent by relatives and friends at home. Shorebound brothers of sailors, perhaps not surprisingly, readily offered unsettling information. Charles Babcock wrote to his whaleman brother Henry in 1845 that Sarah had "proved herself unworthy of you." He added, in true sibling fashion, that "in my opinion the best of girls would not remain true in the absence of 2 years of a lover, mark my word they are faithless things unless you are with them everyday." He kindly offered his brother some difficult advice: "The only way is to do as I do live in the same house with them and then you are shure of her. I would not trust a woman that I loved out of my sight a week I have a very poor opinion of their constancy unless always with them." William Davol gave his seafaring brother Edward some hard lessons of the same sort. Not only did William report that one Eli P. Lawton had recently "waited" on Edward's "June meeting gal" and that if Edward did not look out he would "lose her sure as hell," but he also informed Edward that he and another brother had called off Edward's engagement to "Fallee Brown" because she was not "virtuous" enough. He hoped that the sailor would understand that they had acted in his best interest and that he would not heap "reproaches and execrations" on their heads.[43]

A recurrent dream for anxious whalemen was the homecoming, in which, more often than not, the sweetheart or wife turned away from the newly arrived mariner and gave him the cold shoulder. Alternatively, she asked him to sleep not in her bed but in a less inviting spot. Or, worst of all, she was nowhere to be found when the seaman knocked on her door. Upon inquiring, he was told

[42] Diary of Samuel Braley, *Arab*, date in question (1852?), KWM; diary of Marshall Keith, *Edith May*, 4 April 1868, Log 372, WMLOD.

[43] Charles S. Babcock to Henry Babcock, 28 September 1845, "Cory Family Papers," Mss 80, S–g 3, S–K, S–s 4, Fol. 2, WMLOD; William Davol to Edward Davol 14 May 1847, "Cory Family Papers," Mss 80, S–g 3, S–K, S–s 4, Fol. 7, WMLOD.

she had taken another lover and had moved.[44] Such dreams were founded not on fear alone but on sad experience. Benjamin Boodry would never forget a voyage around Cape Horn when his vessel was shipwrecked, but it was his homecoming that was etched most indelibly in his mind. He learned when he arrived that his lover had been unfaithful. "I lost the ship and lost the girl," he said, "and it worked on me like a dose of phisic." George Blanchard suffered similar misery. Aboard the *America* in the early 1840s, he pined after "Miss Mary." He inscribed a poem of devotion to her, which ended with the following verses:

> Think of those sweet and happy hours
> That now are past and gone,
> When you vowed you would be ever mine
> My beautifull my own!
>
> To find thee true, would well repay,
> The toils and sorrows I have known,
> If fortune smiles, I'll claim thy Hand,
> My Beautiful My own.[45]

George Blanchard arrived in New York to claim possession of Miss Mary in 1842 but was rudely surprised. Written below the last verse of his poem, in what is clearly a later entry, is Blanchard's comment on subsequent developments:

> Fortune did smile and I got Home and found her *darn her* married. Woman, woman, verily thou art the Devil!!

But sadness must have merged with anger for Blanchard, for beneath the exclamations was a small round spot on the diary page. Whether it was a drop of spilled ink or oil or grease we do not know. Blanchard, however, identified it as "*A Lover's Tear.*"[46]

Brothers at home recommended living with women to keep them true. Mariners could do little to follow such advice, but a few found a similar solution closer to hand. Ashore in foreign ports, whalemen might rectify problems of constancy and control by purchasing loyalty with sex. At least a dozen mariners in this sample married prostitutes temporarily. "Dissmissed our wives," wrote boatsteerer Joseph Ray as the *Edward Cary* weighed anchor. "Two marriages took place this day between 2 of our crew and 2 beautys of this country. I wish them much joy," announced Holden Willcox in New Zealand in 1838. One

---

[44] Whalemen who had unhappy dreams of reunion include Marshall Keith, *Brewster*, 10 May 1864, PPL; idem., *Edith May*, 14 May 1868, Log 372, WMLOD; Ambrose Bates, *Euphrates*, 22 May 1860, KWM; Samuel Braley, *Arab*, 10 September 1852, KWM.

[45] Boodry diary, 12 November 1852; Diary of George Blanchard, *America*, 20 December 1841, PC–TB.   [46] Blanchard diary, *America*, undated entry.

week later, a boatsteerer on Willcox's ship was "married for the 3d time since our arrival here; but how long he and his new companion will agree the Lord only knows." The arrangements were all short-lived. On February 19, Willcox reported the following: "All our young men what had got married since our arrival here, got divorced by mutual consent and their late spouses with all their children and donnage either went on shore or on board of some other ships to obtain new employement."[47]

In the light of sailors who "married" prostitutes we might reconsider the familiar sailor's ditty:

> How happy is a sailor's life
> From coast to coast to roam
> In every port to find a wife
> In every land a home.[48]

In addition to regarding this poem as a jab at married sailors who were adulterous or even polygamous, we can also see it as a description of men, mostly single, who attempted to formalize what probably was a casual relationship.

There may have been many reasons why sailors "married" ashore. Perhaps missionaries or natives who shared their beliefs insisted on conjugal relationships. Perhaps the arrangement came at the behest of the "ladies" themselves and served as a means of insuring income or as an expression of emotional attachment. Not surprisingly, this latter explanation is the one favored by sailors themselves. A third mate of the *Bartholomew Gosnold* described the separation of mariners from their shoreside partners as particularly difficult for the women: "The Ladeys go on shore, after biding us all a long adue; and heaving maney A sigh, – for they seem to be very much interested in our wellfare." Women visiting the whaleship *Arab* in Talcahuano harbor felt, according to one seaman, a similar "interest in somebody on board." And when another whaleman "did not stop with [his] wife" because he was not feeling well, he thought "she was rather disappointed."[49]

Clearly, though, the need for "marrying" came in large part from mariners

---

[47] Diary of Joseph Ray, *Edward Cary*, 29 October 1856, NHA; Diary of Holden Willcox, *Columbus*, 10 February 1838; 14 February 1838; 19 February 1838, KWM.

[48] Diary of Isaac Baker, *Warsaw*, 6 April 1841, PMS.

[49] Diary of anonymous mate, *Bartholomew Gosnold*, 29 December 1841, PPL; Diary of William Stetson, *Arab*, 28 February 1854, Log 507, WMLOD; Diary of Ezra Goodnough, *Ann Parry*, 10 June 1847, PMS.

(*Opposite*) Could an American whaleman have conjured up an ideal picture of women at home, it might have looked something like this. This illustration, from *Harper's* magazine, shows two well-appointed Nantucket women gazing out at ships at sea. Research tells us that the women whalemen left at home – wives and sweethearts – were, in reality, too busy, too encumbered by children, or too distracted by land-based suitors to spend many days on rooftops looking oceanward.

themselves. It may have been heightened by sailors' desire for wifely services such as laundry, sewing, or cooking, but it was certainly connected to their sense of the tenuousness of their homebased ties and to their loss of control.

The difficulty was, though, that marriages or constancy, even when purchased from the "ladies," was not to be relied on. Men left brothels for sea just as surely as they had left their hometowns in the first place, and when they sailed out of sight, prostitutes made money with other men. One of the most popular shanteys throughout the age of sail, according to historian Stan Hugill, is one that describes a sailor's unsuccessful attempt to find loyalty among the "ladies." Part of it goes as follows:

> In Amsterdam there lived a maid
> Mark well what I do say!
> In Amsterdam there lived a maid
> An she wuz mistress of her trade
> We'll go no more a rovin' with you fair maid . . .
>
> She swore that she'd be true to me,
> But spent me pay-day fast and free.
> In three weeks' time I wuz badly bent,
> Then off to sea I sadly went . . .
>
> Now when I got back home from sea,
> A soger had her on his knee.[50]

Whaleman Ezra Goodnough knew all too well what was happening in this shantey. Goodnough, who cruised on the Indian Ocean in 1847, explained in his diary how much he needed female loyalty and how hard it was to find, anywhere. If his vessel did not see whales, he said, he would not have enough money to get his girl in Mahe a new dress. And if he did not get her a new dress, "she will forget me." "You see," he explained, "we can hire the girls in Mahe to remember us that is more than the girls at home will do."[51]

Justin Martin, a whaling veteran, had some advice for men like Goodnough. In a letter to a brother who hoped to go to sea, Martin issued a stern warning: "No matter what arrangements you have made," he said, "take a brothers advise and stick fast to old terre firma . . . why you wont get a sight at anything in the shape of a woman the whole voyage and the first one you see when you come back will look so pretty that you will bite a piece out of her cheek before you know it." He urged his brother to remain where he was, to "find a good woman," and to do what other men, ashore and afloat, might also have recommended – to "cage her up."[52]

[50] Stan Hugill, *Shanties from the Seven Seas* (London: Routledge & Kegan Paul, 1961), pp. 46–7.   [51] Goodnough diary, 8 July 1847.

[52] Justin Martin to Charles H. Martin, 29 November 1844, MSM.

Justin Martin points to what seemed to many sailors the most serious social consequences of the whaling voyage. Going to sea elevated a man's hardiness and may have contributed to his manly duty to provide. But it may also have made men overeager for women's company, and it certainly gave them disconcerting evidence of women's potential for independence.

III

The whaling ship was the exclusive domain of working men. This meant that although all sailors performed "men's" work, some of them were also responsible for such jobs as cleaning, washing, and mending that were traditionally associated with women. As one mariner put the matter: at sea there were no "mosquitoes to molest us, no flies to annoy us, and *no women to wash* our dirty clothes."[53] A necessary evil, according to this sailor, was sadly missing.

The delegation of "female" work to sailors may have confounded traditional gender roles, but it seems to have had little effect on sailors' sexual identities. In fact, whalemen minimized in several ways the degree to which they actually crossed conventional gender boundaries. They relabeled the female tasks they performed with male nomenclature. They focused upon and drew pride from the male jobs that they performed at sea in addition to domestic work. And sailors who had the correct credentials – if they were, in other words, white Americans with some spending money – handed women's work over to sailors who were less advantaged.

Officers in the aftercabin, as we have seen, did not have to deal with role improprieties. The captain and officers of the ship had a housekeeper – the steward – and all men at sea were served by a man who took the role of cook. Service positions were frequently filled by black men – usually African Americans – although sometimes cabin boys served temporarily. Not surprisingly, these were men who had limited access to adult "male" roles on shore.[54]

Service workers not only conducted women's work at sea; they were usually delegated separate living space. The cook and the ship's steward were often segregated from men fore and aft and lived in the steerage, located amidships. Their job mobility was impressively nonexistent. Although their wages were not lower than those of skilled seamen (in fact they were sometimes higher),

---

[53] Diary of Isaac Baker, *John Caskie*, 20 August 1855, PC.

[54] James Oliver Horton, "Freedom's Yoke: Gender Conventions among Antebellum Free Blacks," *Feminist Studies* 12 (Spring 1986): 58–9; Leon Litwack, *North of Slavery* (Chicago: University of Chicago Press, 1961), p. 174. Jeffrey Bolster claims that in the early national period African American men worked as sailors aboard ship but in the racially tense period before the Civil War they were often relegated to jobs as cook and steward. See W. Jeffrey Bolster, "'To Feel like a Man': Black Seamen in the Northern States, 1800–1860," *Journal of American History* 76 (March 1990): 1194–95.

cooks and stewards did not overlap with other seamen in job skills.[55] They were not instructed in sail handling, boat pulling, or helmsmanship. On some occasions they were required to assist in an all-hands call, but they were not part of watch duty and they slept through the night. Left out of the power politics that ran from the afterdeck to the forecastle, they were, like the women whose work they performed, excluded from the dynamics of government.

Sailors who were recruited to fill domestic jobs due to a discharge or desertion not only avoided the work but also denigrated those who performed it.[56] They reserved special animosity for black cooks. Sailors on the *Lucy Ann* put gunpowder in the galley stove in 1842 and, according to journal-keeper John Martin, "play all sorts of tricks on the old cook." They were no kinder to his successor and ran "all sorts of rigs upon him." On the morning of July 7, 1843, for instance, the cook was told that one of the "coloured men" was having a fit in the forecastle. When he went to see him, "the scuttle was closed & the deck lights were covered over which made it so dark that he could see no one. They that were below commenced whetting their knives & swearing all sorts of oaths that they had him safe & were going to make sea pie of him, the poor Doc was terribly frightened & begged they would not hurt him as he came down with no evil intention." The pranksters eventually let him out but not until he was "covered from head to foot with flour & chalk."[57]

It is difficult to categorize the frequent enmity directed toward cooks and stewards aboard whaling ships. Much of the violence these men suffered was really directed at the shipmaster or owner, who ultimately controlled the ship's living conditions. But certainly many of these domestics were also damned by the darkness of their skin and by the simple fact that they did work usually reserved for the opposite sex. Comments made by a passenger on a deepwater merchant ship demonstrate the ways that several prejudices may have been intertwined. "You must know," wrote Larkin Turner in 1832, that "we here, in our house, have a kitchen girl – Oh! but this said person is a black man! No

---

[55] Whaling cooks and stewards tended to earn as much as or slightly more than skilled seamen at midcentury. On merchant vessels in the mid-1850s, cooks made $10.00 more per month than skilled seamen. See Lance E. Davis, Robert E. Gallman, Teresa D. Hutchins, "Risk Sharing, Crew Quality, Labor Shares and Wages in the Nineteenth Century American Whaling Industry," *NBER Working Paper Series* 13 (May 1990): 61–4.

[56] A black seaman aboard the *Morrison* objected strongly when he was called aft to take on the duties of a cook who was sick. Merchant seaman William Townsend, called in to handle the galley aboard the *Imaum*, insisted that it was a "berth which I will throw off as soon as possible." Likewise Henry Davis, after a stint as cook in 1862, laughed bitterly at the idea of repeating the experience: "He [the captain] wants me to go Cook as if I was going Cook again now." See the diary of anonymous, *Morrison*, 13 November 1844, MSM; Diary of William Townsend, *Imaum*, 1 May 1858, MSM; Diary of Henry Davis, anonymous merchant vessel, 15 September 1862, MSM.

[57] Diary of John Martin, *Lucy Ann*, 3 July 1842; 7 July 1843, KWM.

matter he answers our purpose ... a red-shirted, raged pantalooned, holed-stocking-heeled, laughing shoed, be dirted, black faced rascal."[58]

Men who did women's work at sea may have been scorned by shipmates, but that does not mean that they internalized the degradation. The diary of an African American steward aboard a deepwater merchant vessel suggests the ways that a man might shape a positive masculine identity from his service work. Charles Benson, from Salem, Massachusetts, sailed aboard an East Indian trader called the *Glide* in the second half of the nineteenth century. He performed work aboard that ship that was very much like the work his wife did at home. He washed floors, he got things "straight & clean" in his pantry, he baked apple pies, and he washed clothing for other men, including undergarments.[59]

Charles Benson was proud that he could do domestic work as well as his wife did. He was pleased with the way he organized the ship's stores and cleaned the cabins, and of the way that he took care of the captain and mate when they became unwell. He liked to think sometimes that he was even better at some of these tasks than his spouse was. "Ha ha," he laughed to himself one day after he had finished washing the cabin, "I should have liked to have you try your hand at it today. I think you would rather scrub your kitchen floor."[60]

Charles Benson cheerfully saw the reflection of his wife in his domestic tasks, but he also drew clear distinctions between his work at sea and her work at home and claimed a conventional male role for himself.[61] Even though he was not trained as an able-bodied sailor and admitted that he had "little to do" with seamen, his self-image was fully bound up with sailing. He began entries in his diary, which was directed to his wife, with sailor talk – with wind, course, and the ship's current position. And he claimed that wild, foul weather, which affected him as much as any "real" mariner, made him both a sailor and a "man." His wife, he said, would never understand what a storm at sea was like. When a gale blew in on a winter's night in 1880, he shared that experience with his shipmates, not his shoreside spouse. "Folks at home do not for a moment think," he remarked, "what a seafareing man has to go through."[62]

The actual distance that Benson kept from a snug home and the simple fact of his seafaring also marked him as unwomanly. "Comfort was there any comfort any time, or anywhere at sea?" Benson remarked. "I cannot rearly say I ever did [see any]. It is the excitement, danger and money that a sea life Brings that keeps me at sea, nothing else." Seafaring, as Benson implies here, also gave him that hallmark of standard masculinity: providership. For all that he performed female tasks, he fulfilled an acceptable masculine prerogative. Benson had left at

[58] Diary of Larkin Turner, *Palestine*, 16 February 1832, MSM.
[59] Diary of Charles Benson, *Glide*, 24 May 1862; 2 May 1862; 6 June 1862; 23 October 1862; 16 November 1864, EI.   [60] Benson diary, 24 May 1862.
[61] On gender ideals among free blacks in the nineteenth century, see Horton, "Gender Conventions."   [62] Benson diary, 14 May 1862; 11 December 1880.

home at least two children with his wife, Jenny, and his reasons for seagoing were tied closely to his desire to support them. "What a miserable life a Sea fareing life is," he lamented in May 1862. "I will stop it if I live & that soon . . . [but] you & the children must have things to eat drink & [wear] & I must get it some were if not on land, on the sea."[63]

Stewards and cooks were not the only men on shipboard delegated women's work and not the only men to take a positive approach to their jobs. Sailors before the mast performed ship's duty and whaling work as part of their required labors, but they also had to do their own sewing and washing. They did not universally find these tasks distasteful. Francis Moreland aboard the *Sooloo* in 1861 had a "grand wash" with his shipmates. James Haviland, like other sailors, saved his washing until he could stand it no longer, but he then claimed with pride that he could "do it as quick and as well as any washwoman." Robert Weir was pleased to be able to make himself a patchwork cap and thought of those at home who were more familiar with such work: "How often do I think of my dear sisters while engaged upon all this kind of work wouldn't they laugh and criticize – but they may yet find out I am not so poor a knight of the needle after all."[64]

Foremast hands took special satisfaction, not surprisingly, in domestic duties that could be performed communally. "Started the sewing society again," noted Robert Weir, "stitch on stitch, patch on patch is all the rage." Sailors aboard the *Illinois* had a regular quilting bee. "One of our fellows quilt wanted quilting," reported Elias Trotter, "He got 4 Iron poles for a frame & adjusting it in rather an odd manner, called upon his watch for assistance. The jacks both old & small, brought their needles & their palms & sitting around on chests & buckets & barges, surrounded the quilting by the number of ten – Twas an odd sight to see these 'heavy dogs' use the needle & quilt – some smoking others chewing, while in rotation they spun their yarns."[65]

Recent research on the delegation of work in antebellum America suggests why many of these sailors did not balk at occasional domestic work. Within farming families in the northeastern United States, men not infrequently assisted women in the home. Although house work was primarily defined as women's labor, men were not unused to domestic jobs. This contrasted with the gender roles among the urban middle class, where there was a more absolute division of labor.[66]

[63] Ibid., 11 August 1880; 13 May 1862.
[64] Diary of Francis Moreland, *Sooloo*, 12 October 1861, PMS; Diary of James Haviland, *Baltic*, 26 January 1857, PPL; Diary of Robert Weir, *Clara Bell*, 16 April 1856, MSM.
[65] Weir diary, 10 April 1856; Diary of Elias Trotter, *Illinois*, 3 December 1846, Log 1005 A&B, WMLOD.
[66] Nancy Grey Osterud, *Bonds of Community: The Lives of Farm Women in Nineteenth-Century New York* (Ithaca: Cornell University Press, 1991), pp. 139–86; Karen V. Hansen, "'Helped

All this is not to say that whalemen relished role reversals. Many men before the mast sought to translate their domestic duties into tradesman's work. Sewing societies became men's workshops, for example. Henry Davis explained to his diary reader that if anyone looked into his forecastle "one would think it was a Tailor's shop for one is making Pants another Shirts another is puzzling his brains over a Cap and the last Knight of the Needle is mending."[67]

Affluent seamen were particularly successful at distancing themselves from domestic tasks in part because they had the means to avoid them. Robert Weir hired "Manuel" to wash his clothing. Lewis Williams paid "Antone" a pair of pants to wash his clothes, and Elias Trotter made a bargain with "Valentine Millet (a Colored man) that He should do my washing, mending, & shaving during the whole voyage for 17 dollars."[68] These excerpts indicate that sailors of color were again the most likely individuals to take on women's work. Performing only men's work was a luxury not all men could afford.

Most young sailors hoped that the whaling voyage would set them back on shore with socially marketable skills and with independent means. As the preceding discussions suggest, however, whalemen faced special challenges when it came to gaining masculine stature. Their proximity to shipmasters' wives limited their efforts at independence. Their distance from hometown sweethearts threatened their bargaining power in courtship. Then, too, men at sea had to perform women's work, a social obstacle that some men overcame with ease, others by imagination or money. Throughout the voyage young men held tight to a sense of themselves as distinctly, even strikingly, unwomanly. This was despite the many difficulties of becoming a man at sea, but it was also because of them.

IV

Understanding the extent to which seafaring men were or became distinctive from landsmen, either in identity or behavior, becomes especially perplexing with regard to sexual or romantic practice on shipboard. The foregoing discussion centered on whalemen's relationships with women, both fictive and real, in large part because most available evidence concerns relationships between the sexes, with little mention of male romance or homoerotic behavior. It is thus impossible to treat adequately the latter subjects, but the small amount of data

Put in a Quilt': Men's Work and Male Intimacy in Nineteenth-Century New England," *Gender and Society* 3 (September 1989): 334–54.

[67] Diary of Henry Davis, anonymous vessel, 26 May 1862, MSM.

[68] Weir diary, 18 April 1858; Diary of Lewis Williams, *Gratitude*, 29 March 1859, PMS; Trotter diary, 22 August 1845.

that has surfaced is worth examining, if for no other reason than that it emphasizes the need for further investigative work.

The issue of sexual contact between men on shipboard arises in the material examined here (and this survey includes court records and more than 1,000 ships' logs) only three or four times. The cases are so briefly mentioned that each can be stated in some detail.

In early September 1838, according to shipmaster Sanford Wilbur, the cabin boy of the whaleship *Joseph Starbuck* of Nantucket reported to the master that the third mate "had Conducted himSelf verry strangely the night Before & he Recuested me to let him exchange watches." The captain refused to let him do so but asked the boy to let him know of any untoward events. "At 11 A M," he reported, the cabin boy "came Down from the helm and Informed Me that [the third mate's] conduck towards him was as it had Been a Number of times Before Such as trying to kiss him & caress him." The boy told the officer to let him alone, the captain noted, "But he still persisted he than Saw him take Something out of his Pocket & herd It open and he supposed It to Be A Raizer than [he] cauld upon the Boy & wanted him to Do Something for him." The boy resisted and ran to find the shipmaster. Within a short time the "Cabin was alarmed & all hands turnd Out & we all Could not account for his conduct whether he ment to take his life or the Boys or to frighten the Boy Into Some Heighneous Crime." When ordered into the cabin and questioned, the third mate protested that the accusations were a "Damed ly," and after an altercation he was confined in irons and a guard set over him. "My officers and Cabin Company," concluded Wilbur, "then one and all Protested against him & his Conduck and wished for me to keep him in Clost Confinement & put him on Shore as fast as Possible which it is My Intention to Do." Eight days later, the third mate was set alone on shore on the nearest landfall – the Chilean coast.[69]

An incident aboard the whaleship *Emily Morgan* in the spring of 1835 is similar in important ways to the *Starbuck* event. According to testimony signed by twelve foremast hands and directed to the American consul at Paita, Peru, a sailor named John Fryer had, besides stealing, deserting ship, and fighting, been guilty of "trying to have connection with David Stanton in his hindemost parts" at least twice. Fryer, "fearful of the punishment," took "4 spoonfuls of spirits turpentine and quicksilver to put an end to himself" and "stowed him self away under one of the lower berths in the forecastle." When his shipmates discovered him (and apparently did something to effect his recovery), he threatened them with a knife. The following day, everyone on the vessel – all officers and crew – "came aft and said it was not save to have sutch a man on board of the ship and requested that he should be put on shore."[70]

---

[69] Logbook of master, *Joseph Starbuck*, 7, 15 September 1838, NHA.

[70] George Ray, Note to Paita Consul, 21 May 1835, in Crew List of *Emily Morgan*, NBFPL.

In a third event, the master of the ship *Citizen*, Hiram Bailey, wrote to his owners of the "misdeeds" of the ship's steward. Three of the "green boys" of the forecastle reported to the captain that the steward had entered the forecastle at night, "got into there berths when the lights were out and took there inexpressibles in to his mouth." The captain planned to remove the steward from the ship the following day, when the vessel reached Faial.[71]

Finally, two sailors of the ship *Montreal* brought suit against a shipmaster named Nathaniel Soule in 1861 for sodomy. Their case was heard in the admiralty court in Honolulu, and the decision of the Court in both (separate) cases was against the defendant. He was ordered to pay the sum of $1,500 and costs in one instance and $2,500 and costs in another. Captain Soule claimed that he had been framed. "I trust yet to show," he remarked, "that I have been the victim of a conspiracy, which, to use the language of the learned judge, 'the annals of criminal jurisprudence scarce contains.' "[72]

Evidence like that just described obviously raises as many questions as it answers. Why, we wonder, do the few cases that have surfaced in thousands of ships' logs and legal records describe only coercive acts between men? Does this suggest that sexual contact between consenting sailors was not considered criminal at sea or worthy of remark, or does it suggest that it did not regularly occur? And to what extent does the paltry evidence reflect not incidence but simply the reluctance of Victorian notetakers? Some recordkeepers were clearly hesistant to note specifics of alleged transgressions. We know very little, for instance, about "A disturbance" that took place aboard the ship *Meridian* when a "Canaker boy" was accused of doing "A thing that he had no business to Do to John Ferry" – only that John Ferry was considered the guilty party and whipped. We also do not know exactly what prompted the entire crew of the *Charles Phelps* to bear witness against the steward, William Smith. Although the journal-keeper was willing to inform his readers of the length of the knife that the steward was alleged to have carried (12–13 inches long) and went into other minute detail, he was loath to note the seventh charge against the steward. He would say only that the sailor was accused of "gratifying the carnal and craving desir," and that "modesty and decencey compel me not to give the particulars of this act."[73] If modesty and decency prompted censorship here, even when details might have been necessary to justify severe punishment, so much more might modesty veil less consequential actions, as well as actions that involved mutual consent.

It is difficult to assess incidence through selected reports of "criminal"

---

[71] Hiram Bailey to Charles G. Coffin and Henry Coffin, 15 September 1844, collection 152, NHA.

[72] "Whalemen's Shipping List and Merchant's Transcript," 12 February 1861.

[73] Diary of Henry W. Collins, *Meridian*, 10 January 1830, PPL; Diary of Silas Fitch, *Charles Phelps*, 29–31 January 1843, MSM.

behaviors, and it is similarly hard to elicit shipboard attitudes from these
accounts. The punishments described here – in several cases involving the immediate ejection of the sailor from the ship's crew – were, by any comparative
measure, severe. Yet, before we can conclude that sailors both fore or aft were
intolerant of attempted sexual contact at sea, we must consider extenuating
circumstances. In almost every case, the criminal charge involved complicating
issues of class and race. The steward of the *Charles Phelps*, for example, had,
according to diarist Silas Fitch, several "liabilities" other than his alleged im-
proper behavior. "He is," said Fitch, "the proudest and the sasyest darkey that
I ever saw he is not only proud and sasey but he is dirty and lazy he profeses
to be A roman catholic and I think he is for I have often herd him curse and
sware and that I think is their creed according to what I know about them."[74]
Intolerance aboard this vessel may have come in several guises. Furthermore,
the willingness of both foremast hands and officers to join together in chastising
an offender from the aftercabin may have stemmed from the forecastle's interest in turning political tables on accused officers.

Almost all the foregoing evidence also describes events involving perceived
threats to the safety of the crew. The accused men were said to have threatened
their shipmates with knives, razors, and poisoned food. A clear reason, then, for
severe and immediate punitive action might have been the security of the ship.
Sexual conduct might have had little to do with the situation.

Looking at the matter in another light, however, suggests that sexual conduct
may well have been what provoked severe punishment. Sailors who were charged
with threatening with weapons or assaulting officers (both of which constituted
mutiny), and who were not charged with sex crimes were frequently kept on as
part of a ship's crew after they were disciplined. On the ship *St. George*, for
instance, a forecastle hand who hit a third mate on the head with a handspike
was put into irons and confined in the ship's run but was later released into the
active workforce. Another man who threatened the ship's captain with a sharp
knife was incarcerated for several weeks and then sent forward. On the bark
Sarah in 1866, a sailor who wielded a knife and warned that he would kill the
crew and officers and "smash their brains out" was confined for three weeks but
then was released when he promised to behave.[75] The officers on these ships
may have been particularly accommodating because they lacked alternative
manpower, but these are not isolated cases. The instances in which sailors were
hurriedly ejected from a ship, especially through the insistence of forward
hands and the afterguard, were rare.

Perhaps, then, many sailors had difficulty accepting sexual intimacy at sea.

[74] Fitch diary, 11 January 1843.
[75] Diary of John Martin, *St. George*, 21 February 1866; May 1867, PPL; Diary of Amos Baker
[?], *Sarah*, 28 June, 19 July 1866, PPL.

When John Fryer learned that his attempts at sexual contact had been found out, he drank mercury and turpentine to commit suicide. His reaction may suggest intolerance for sexual acts among a ship's crew.[76]

Any intolerance that existed, however, would probably have been unevenly expressed from forecastle to afterdeck. Within the forecastle, as we have seen, veteran sailors stressed a public commitment to the group as a whole. Sailor couples as well as sexual pairs may have been antithetical to the fraternal ethos of that community.[77] Sailors who sought each other out for romantic friendships suggest that such was the case for them. These sailors, of middle- or upper-class backgrounds, were like men ashore who carried out intense, intimate, and sentimental relationships with one another. At sea, they sought and occasionally found a compatible soul. Elias Trotter relied on gams to meet men of his sort. At the end of November 1845 he found a sailor, Charles Wheeler, who "drew [his] attention" for his "manly beauty, activity, and intelligence." They carried on secret conversations for the course of the gam but then had to part. As Wheeler's ship sailed astern, Trotter realized that "in all probability they would never meet again." On his own ship, though, he discovered a man who would be his intimate. He was a sailor named Longworth, who was a man of "some education, . . . sensitiveness & discernment." In the night watches and on lookout, Trotter tightened his bond with this man. We do not know whether their romance extended to sexual intimacy, but Trotter suggests that neither physical nor emotional coupling was possible for anyone within the forecastle. "It is not natural," Trotter complained, "so many beings imprisoned in a small compass & living with the eyes of each continually fixed upon another."[78]

Intimacy of any sort may thus have been proscribed by most communities of men before the mast. The utter lack of privacy, the proscription of exclusive behavior, and the social prejudices that men may have carried aboard all may have made forecastle hands firmly unwilling to countenance sexual relations.

For their part, ships' officers stood officially against sexual contact at sea and delivered severe punishments to sexual offenders. They may have acted out of

[76] Note of George Ray, *Emily Morgan*, 21 May 1835.

[77] Karen Lystra maintains that Victorian lovers built unique bonds that "emphasized their individuality, their distinctiveness, and their separateness." See *Searching the Heart: Women, Men, and Romantic Love in Nineteenth-Century America* (New York: Oxford University Press, 1989), p. 9.

[78] Trotter diary, 26 August 1845; 29 November 1845; 14 December 1845. Romantic friendships on shore, suggest E. Anthony Rotundo and Donald Yacovone, did not necessarily involve sexual contact. See Rotundo, "American Manhood," pp. 75–91, and Yacovone, "Abolitionists and the 'Language of Fraternal Love,'" pp. 85–95. Other important work on male friendship in this period includes Robert K. Martin, "Knights-Errant and Gothic Seducers: The Representation of Male Friendship in Mid-Nineteenth-Century America," in Martin Duberman, Martha Vicinus, and George Chauncey, Jr., eds., *Hidden from History: Reclaiming the Gay & Lesbian Past* (New York: Meridian, 1990), pp. 169–82.

a sincere belief in the sinfulness and criminality of homoerotic acts. But if they felt physically attracted to sailors themselves, so much the better to discipline a subordinate – as a way of deflecting suspicion away from the aftercabin or even as an act of self-chastisement. Arthur Gilbert, in a study of buggery in the British navy, has asserted that the "ferocity" with which authorities punished men for sodomy in the eighteenth and nineteenth centuries was in part due to the navy's nervousness over its reputation.[79]

An irony of the aftercabin was that, although it housed officers formally committed to eliminating sexual acts at sea, it was the one place where physical contact might easily take place. Behind closed cabin doors, men had the freedom and the privacy to meet intimately. As long as these meetings reflected mutual interest and were discreet, there was little risk of public exposure. And even if they were not mutual – the proverbial cabin boy and captain encounters come to mind – ships' officers who controlled the vessel's logbook could delete any unbecoming conduct from the official voyage record.

One can easily conjecture about male sexual practice on whaling vessels, but solid evidence is more elusive. Until we know more about sexual expression at sea, particularly about noncoercive contact, we can only speculate about how sailors met physical needs and conformed to or veered from seafaring and land-based norms.[80]

---

[79] Arthur N. Gilbert, "Buggery and the British Navy, 1700–1861," *Journal of Social History* 10 (1976–1977): 72–98; for the U.S. Navy see also Christopher McKee, *A Gentlemanly and Honorable Profession: The Creation of the U.S. Naval Officer Corps, 1794–1815* (Annapolis: Naval Institute Press, 1990), pp. 437–42.

[80] The subject is beginning to open up. George Chauncey, Jr., has done important work on sexual identity and sexual behavior among working-class sailors in the twentieth century. See George Chauncey, Jr., "Christian Brotherhood or Sexual Perversion? Homosexual Identities and the Construction of Sexual Boundaries in the World War I Era," in Duberman et al., *Hidden from History*, pp. 294–317. B. R. Burg has recently brought to light the confessional diaries of an enlisted man in the U.S. Navy in the mid-nineteenth century. These diaries describe widespread sexual activity in this service. B. R. Burg, *An American Seafarer in the Age of Sail: The Erotic Diaries of Philip C. Van Buskirk 1851–1870* (New Haven: Yale University Press, 1994).

Afterword

I
"All the Honor and the Glory":
The Men Who Went Home

Three to five years passed, and most young whalemen – grown perhaps from their teenage years to beyond the age of maturity – returned to the American mainland. With their faces weatherworn, their clothes a quilt of patches, and their muscles taut with trials of endurance, they carried with them the signs of physical change.

For all their physical distinctiveness, though, the men who disembarked from a whaler were not utterly different from the landsmen they had been when they first set foot on the deck of a ship. They had spent years in a vessel on foreign seas, but in a sense they had sailed in a floating home. The whaleship bore the stamp of the society that launched it. The divisions that it maintained, with places for men of certain ages, races, and ranks, as well as space for "women," reflected the social structure on land. And whaling work was organized by the social and economic interests of men, not as they faced the sea from the bow of a ship but as they envisioned the voyage from the counting house.

Whalemen not only worked in a setting that was strangely familiar; they persisted in viewing seafaring through the lenses of landward cultures and in judging it by standards set on shore. They shaped their maritime behavior according to their imagined place – past or future – back home. They held as tightly as they could to familiar social attitudes, and when ashore in foreign ports, they moved in circles in which they felt comfortable. Even when some of them crossed at sea the most tightly built social boundary of all – the gender division of labor – they endeavored to minimize the change and to preserve the sanctity of their shore-defined maleness.

Yet for all that these men remained landsmen and for all the ties to home that they did not loosen, they had also experienced something unique. All whalemen took part in a venture that employed a distinctive vocabulary and set of skills, and they lived in a desolate setting apart from women and well beyond the landsman's ken. They had together kept a small wooden ship afloat and on course and had faced the daunting task of pursuing whales while staying alive. All this had oriented these men not to past or future interests but to the men around them and to the here and now of shipboard survival.

As they forged a community with their fellow sailors and faced the common foreign element around them, these seamen frequently began to speak of themselves in contradistinction to landsmen. They did not all agree about what made them different. Sometimes it was that they were braver or that they dealt with greater hardship or that they felt life more keenly than did their counterparts ashore. Sometimes it was that they were more appealing to women than landsmen were; other times it seemed that women loved them least of all. But most whalemen – foremast hands, officers, and Victorian outsiders alike – agreed that their sojourn at sea had somehow set them apart.[1] This experience linked them to other Americans who felt that they stood on the outside of society looking in. There were many men and women who, by virtue of their birthplace, their race, or their lack of money, felt bypassed by the dominant American mainstream. After years at sea, deepwater sailors, no matter how high born, shared their sense of estrangement.

It had been within the forecastle particularly that American whalemen had experienced something of a sea change. The confines of these living quarters and the isolation of the ship had forced sailors to yield some of their shore-born loyalties and to make social peace with men of different origins and upbringing. Hard treatment from the afterguard further demanded that foremast hands suspend divisions and unite in protest. Their earnings, their health, and their general welfare had depended upon the outspoken strength of their fraternity.

Because most of the sources on which this study is based stop at the voyage's close, we do not know how many whalemen were permanently affected by the forecastle community, the ship's company, or the sense of a life apart. Historians tell us that merchant seamen in the late 1700s carried with them an imprint of work afloat. They applied the lessons of the forecastle to politics and social relations ashore and were radically influential among working people in seaport settings.[2] Whalemen before the mast certainly were schooled in antiauthoritarian protest and collaboration, but there was enough resistance to

---

[1] References to sailors as distinct from landsmen are numerous. See, for example, Diary of Edwin Pulver, *Columbus*, 9 January 1852, PPL; Diary of Abram Briggs, *Eliza Adams*, 30 May 1873, Log 940, WMLOD. Herman Melville uses the same set of oppositions in *Moby Dick* (New York, 1851; reprint: New York: Norton, 1967), p. 12.

[2] See footnote 12 in the Introduction.

collective action at sea to suggest that many may have arrived home and dispersed quickly to separate shoreside niches. There were many whalemen, too, who came from rural inland areas and who may have been only transient figures on the waterfront. Once a voyage was over, these men may have left port and ships and sailing for good.[3]

Conclusions about the impact of seafaring experience on whalemen and other sailors await shore-oriented studies. There exists evidence, however, that at least one former whaleman carried the lessons of the forecastle and the ship well beyond the voyage itself. Herman Melville, experienced foremast hand, devoted much of his work to celebrating male company, to honoring "Otherness," and to questioning autocratic power. In all his novels, but particularly in his work about whaling, he extols those men who cast aside shore-based prejudices and embraced social equality and community. His heroes are men who are truly able, in the terms of King Neptune, to "cross the line."

The sources of Melville's inspiration were numerous, but whaling, it seems, had a formative influence on the author's genius. He made this explicit through the character of Ishmael in *Moby Dick*. In one of that novel's best-known passages, he indicated how much he owed to his experience in the fishery:

> If, by any possibility, there be any as yet undiscovered prime thing in me; if I shall ever deserve any real repute in that small but high hushed world which I might not be unreasonably ambitious of; if hereafter I shall do anything that, upon the whole, a man might rather have done than to have left undone; if, at my death, my executors, or more properly my creditors, find any precious MSS. in my desk, then here I prospectively ascribe all the honor and the glory to whaling; for a whale-ship was my Yale College and my Harvard.[4]

Herman Melville's whaling novel did not find ready acceptance among gen- teel readers in the nineteenth century. The author's eclectic approach and his celebration of the "uncivilized" did not sit particularly well with a social estab- lishment that extolled the virtues of rank, separation, and self-involvement. His "mutual, joint-stock" vision of things was also out of step with men who centered their lives around the importance of competitive individualism and who spent much of their time not breaking down social boundaries but build- ing them up. Melville, like whalemen and sailors in general, had to wait for another century before the general reading public and the academic community

---

[3] Whaling crew lists, which show a continuously youthful population before the mast, indi- cate that the majority of men from the forecastle left whaling either during or at the end of their voyages. Whether they abandoned seafaring altogether or went, for instance, into the merchant service or navy, we cannot easily know. See Appendix II.

[4] Melville, *Moby Dick*, p. 101. My assumptions about Melville's other work are based on readings of *Benito Cereno*, *White-jacket*, *Billy Budd*, and *Redburn*.

celebrated, rather than denigrated, their vision, their labors, and the shape of their lives.[5]

II
"The spell that binds us": The men who sailed on

... but still we go, leaving wife, children, home, everything in the dim distance. We go forth to battle with the deep and its monsters ... we are not satisfied with dry land, with its smiling bounties and lovly women, but must rush headlong into we know not what, until it is too late, Old Neptune has had his hand upon us, and we are doomed to a life on the ocean wave, without we have the moral courage to shake off the spell that binds us to the sea.

Ascribed to John Beebe, *Tropic Bird*, c. 1850s (NHA)

The whaling voyage was ideally a rite of passage. The returning ship delivered to shore a man fully grown who disembarked to buy his farm and to find his sweetheart waiting. He left the sea and all the promises it made to other young and needy individuals. It did not always work that way, of course. Many whalemen returned home to claim more of a pittance than a prize and to discover that their sweethearts liked suitors who were more attentive. And some men did not return at all. At least one man died on each of the voyages described in this account. Lost in shipboard accidents, killed on the whaling battleground, or laid low by deadly diseases, these young men never knew the joys or the disappointments of homecoming.

There were, too, those seamen who went home, married, made love, kissed their wives (and babies), and set sail again. And again. Samuel Braley, devoted husband to Mary Braley, claimed that whaling was the "most disagreeable of all professions," yet he remained at sea until he died of dysentery in 1870. Ambrose Bates, fully enamored of his wife Annie, was equally adamant about disliking whaling: "This is a hard wo[r]ld," he wrote in 1861, "and I believe this to be the most loathsom avocation therein." Bates, though, made no plans to leave the sea.[6]

The men who repeatedly turned their backs on home, family, and friends were in some ways the true dissenters from the American mainstream. Yet

[5] Historians and sociologists describe how, toward the turn of the century, with the rise of geographic mobility, a national market economy, and corporate bureaucracies, the rise of "other" direction became culturally widespread. Middle-class Americans found it more and more necessary to be socially flexible and sensitive to others around them. These tendencies, according to an optimistic Jackson Lears, have since "become common in our own time." See T. J. Jackson Lears, *No Place of Grace: Antimodernism and the Transformation of American Culture 1880–1920* (New York: Pantheon, 1981), p. 35.
[6] Diary of Samuel Braley, *Arab*, 14 June 1850, KWM; Log of the *Sea Fox*, 20 August 1870, KWM; Diary of Ambrose Bates, *Nimrod*, 31 March 1861, KWM.

they, too, bridged the expanse from shore to sea, and they confound labels that dismiss them as social outsiders or as anomalies. The shipmaster, who was usually the oldest salt of them all, was, as we have seen, laden with land-based agendas and tethered as much as he was free. And experienced sailors who were married (usually officers) frequently cast their ongoing decisions to sail not in terms of an escape from the confines of home but as an act of domestic obligation. Shipmaster Samuel Braley felt that he "must struggle on, or starve, and not only me but you and the little responsibility too." If it were not for his wife and his son, he claimed, he would "curse the whales and go cod-fishing." Marshall Keith, who was "homesick as a dog," planned to leave the voyage early to go home to his wife Sarah. When he heard she was pregnant, he changed his mind. "I have said . . . that I was coming home from the islands but when I Heard that I was likely to have a new comer I thought it my duty To go the voyage to get something to support it and my little darling." This manly duty was something Keith was proud to carry out but something onerous as well. In the back of his mind stood his wife Sarah, who, he felt, would be disappointed if he did not pursue the voyage. "I do not expect you will give me any welcome," he imagined her saying if he came home early. Bound by duty, therefore, Keith sailed on.[7]

Men who sailed out of love and obligation did so because marital and familial ties were strong, and they paid a painful price in separating from home. Edward N. Jenney, mate on the *Alfred Gibbs*, became depressed and distracted as he left port in November 1859 and was obsessed with thoughts of his wife, his children, and a sister he had left sick. "Oh my God what shall I do," lamented Jenney. "O what I would give too live my life over again and know what I now do. But it is too late now *yes too late*. the Dice is cast. there is no going back no standing still. moove moove is all I can do." Two days later Jenney confessed to his diary that the thought of not seeing his sister again was driving him to distraction. "Oh God, I am near crazy," he declared. Two weeks later he was as despondent as ever: "What shall I do My little children. I wish I had never been born."[8]

Jenney's misery is mirrored by that of Samuel Braley. "Talk about dieing," mourned Braley on the first day of a four-year voyage. "I hope that I shal never suffer any keener pangs than those that now pierce my heart." Two months later, in a more speculative frame of mind, Braley pondered the "exile from all I love" and the "torture" that he felt he endured in a life at sea:

> I often ask myself the question for what are you toiling? for the time will
> soon come when you must leave all and; it will make no diference to you

[7] Braley diary, 14 February 1850, 27 October 1850; Diary of Marshall Keith, *Cape Horn Pigeon*, 1 July, 1 October, 14 August 1866, Log 371, WMLOD.

[8] Diary of Edward N. Jenney, *Alfred Gibbs*, 12, 14; 30 November 1859, Log 998; WMLOD.

wheather you leave much or little; I want to leave just enough to pay the saxton that places the last sod on the grave of my wife after she has been well supported through life; and not a dollar more.[9]

Some shipmasters had ways of mitigating their desolation. They persuaded shipowners to allow them to sail with their spouses. Or they promoted social programs, often centered around piety and temperance, that their wives would have appreciated. Other sea captains created a space aboard ship that stood for a woman's presence. They transformed the aftercabin into a home of sorts, complete with a parlor, a pantry, and a bedroom, all furnished with domestic items.

Officers' rituals also acknowledged – even if in limited ways – the importance of sweethearts and wives. Not only did the Saturday night ritual honor women left behind but also the consumption of "home cakes" comforted men at sea. Home cakes were tasty concoctions that spouses or special friends sent aboard a ship at departure, intending that they be eaten at a later date. They were so well preserved with alcohol that opening one was sometimes a hazardous enterprise. Marshall Keith's home cake, made by Sarah Pope, his betrothed, was so explosive by the time he opened it five months after sailing that he "nocked one of them higher than a kite." It was "in good order," though, and he relished it.[10]

Marshall Keith honored his fiancée's birthday with his home cake and thus, with her help, collapsed the gap between ship and shore. George Bowman, a married second mate on the *Albion*, was also symbolically carried homeward on his anniversary by a box of cake. Captain Edmund Jennings saved his home cake for Thanksgiving day, but then the comparison of his lonely circumstances with the imagined happiness of the holiday ashore sent him into a fit of despair: "This is your thanksgiving Day with you how different it is with us a gale of wind, Our Bark tossed about. I cut Aunt Dyers Cake, and gave it to the Steward and then went into my room and had a good crying spell. I wondered if they missed my place at the Table. Or if [they] thought of me. I suppose Some thinks of me on all such occasions. May God bless and keep you all."[11]

As one might expect, there is more to the story of veteran whalemen than domestic duty and unequivocal devotion to women. Many sailors felt deeply divided about permanently staying at home. Samuel Braley, who pined for his wife, also admitted that seafaring was a "pill of his own choosing" and that he had serious doubts about living ashore. He acknowledged that if he "was with her now, one month would make [him] sigh for the sea." Allen Newman aboard the *Covington* in 1854 had something similar to say: "O could I be at home to

---

[9] Braley diary, 22 November 1849; 25, 27 January 1850, KWM.

[10] Diary of Marshall Keith, *Brewster*, 9 February 1864, PPL.

[11] Diary of George Bowman, *Albion*, 21 June 1869, PPL; Diary of Edmund E. Jennings, *Mary and Susan*, 29 November 1877, PPL.

night, I dont know of any inducement that would tempt me to leave before morning, God bless My Wife & Children." [12]

For these men, it seems, a man's duty to provide was not only a reason to be at sea but also an excuse to be there. Shipmasters' wives emphatically confirm this impression.[13] They suggest in diaries and in letters to absent husbands that when duty involved years of separation, it had something of a false ring to it. Duty to them meant not only breadwinning but also assisting on the homefront.

Mary Chipman Lawrence, who sailed with her shipmaster husband aboard the *Addison* in 1857, had something to say on this subject. She met a Captain Daggett at a gam and noted with disapproval that he was "sixty years of age and is taxed for $70,000." Lawrence told him directly that she "thought he was a very foolish man for leaving his family at that age, when he could live comfortably at home." She had similar things to say about a Captain Ludlow, who made a sailing voyage for each of his three children. Despite the obvious fact that Ludlow was attempting to provide each child with material comfort, Mrs. Lawrence remarked that "it is a question in my mind whether it is advisable to do so, whether the children would not be as well off without it."[14]

Mary Chipman Lawrence's sentiments about seafaring fathers and husbands were shared by other women. Susan Cromwell made her priorities patently clear in writing to her husband aboard the *Reindeer*: "May you never again leave me, for your society is more than all you can gain of this worlds goods." Julia Gates hoped that her husband would find work on local steamers. The couple might then be less financially secure, but at least they would be together more frequently. "Oh George," she wrote to her spouse, "how I wish you was in the steamers as you used to be. I think the pleasure of being at home now and then would overbalance all the extra money making. Do you not think so?" George Bowman's wife, whom he suspected was working "beyond her strength" in his absence was willing, he admitted, "to work like a slave if I would only stay at home."[15]

The disjunction between what women wanted from their husbands and what some seafaring husbands wanted for themselves is particularly evident in the long-distance letters that men sent home. In contrast to women on shore who complained of being overburdened and lonely, sailors were frequently cheerful and upbeat. William Allen exhorted his wife Lizzie to "keep up as good courage as possable." Henry Beetle urged his "dear" Eliza to "keep up good heart." Captain Albert Goodwin had a more concrete suggestion for his wife:

[12] Braley diary, 11 March; 12 May 1850; Diary of Allen Newman, *Covington*, 17 February 1854, PPL.
[13] Discussion of shipmasters' wives here is based on the journals and correspondence of twenty wives of masters, all of whom lived in southern New England.
[14] Diary of Mary Chipman Lawrence, *Addison*, 11 March 1857; 27 May 1857 in Stanton Garner, ed., *The Captain's Best Mate* (Providence: Brown University Press, 1966).
[15] Susan Cromwell to Peter Cromwell, 1 January 1854, DCHS; Julia Gates to George Gates, 18 December 1871, MSM; Bowman diary, 25 December 1868; 25 June 1869.

Such a poignant scene, with the departing sailor gesturing expansively and
looking reluctant, and the woman leaning in a posture of sweet resignation,
belies the tense domestic negotiations that often characterized leavetakings,
especially for veteran seamen. This image was one of the most popular
sailor scenes produced by the Victorian press.

Now do not my dear Eliza get yourself down and mourn over your desolate
state, but in the summer take a journey some where. You know how much
better you feel after it; do not think so much of the expense I will take care
of that.[16]

[16] William Allen to Lizzie Allen, 21 July 1864, NBFPL; Henry Beetle to Eliza Beetle,
17 February 1848, PPL; Albert Goodwin to Eliza Goodwin, 28 December 1844, PMS.

In response to their wives' pleas and possibly to their own consciences, veteran sailors promised over and over again that their present voyage would be their last. Jared Gardner asserted in a letter to his wife that on that "happy day" when he would see her again, he would "submit to anything that you requested." What did he imagine his wife would ask for? That he would not leave her again "without it be by the hand of death." Gardner announced to his wife that he was willing to take that step. "I think that I should answer without hesitation I never can nor I never will." But he continued to sail.[17] In a similar vein, Henry Hobart Marchant wrote to his "Affectionate wife" Mary in June 1839 that "let me have the pleasure once more of seeing you & I think we will never part till deth seprates us." But in 1850 Marchant was still on the other side of the globe.[18] Finally, Captain William Jackson vowed to others that he would forgo the sea for the sake of his family. Jackson wrote to a friend in January 1860: "Now here is my dear wife willing for me, not to go again, saying lay aside your Charts and stay at home with me. . . . My Caroline's health is not of the very best, & she does not wish for me to leave her again and my little Daughter is a bright loving little thing, ten years old, their united persuasions are enough almost to persuade the brooks from running toward old Ocean, and I dont see how I ever can go again." These "united persuasions" were just not enough, however, to keep this man ashore. By the following June, William Jackson had left his ailing wife and bright young child for a whaling voyage in Hudson's Bay.[19]

Jared Gardner, Henry Marchant, and William Jackson do not say why they felt bound to a seafaring life, but other sailors were more forthcoming. One whaleman claimed that it was not simply that the ocean enticed men from home, but that the "dull, tame shore" in general and the *untamed* domestic circle in particular pushed them away. To make his point, this diarist used the example of a sea captain named Stetson. Captain Stetson was worth $100,000 "in cold cash." When he first got home from a voyage, he was determined to stay ashore. He "would take his seat in the chimney corner and smoke his pipe." After a while, however, "his return got to be an old story in the house," and the "women began to shove him from one corner to another, and he could not stand it, so he was obliged to come away."[20] It seems clear that some shipmasters, used to a degree of obedience and deference at sea, had trouble dealing with women at home who flexed their domestic muscles.

Many veteran whalemen went beyond issues of domesticity to explain what sent them to sea repeatedly. They claimed that they became career sailors for the same reasons they had gone whaling in the first place: a desire to see more

[17] Jared Gardner to Harriet Gardner, 27 August 1842, "Gardner Family Papers," AAS.
[18] Henry Marchant to Mary Marchant, 20 June 1839; 30 March 1850, DCHS.
[19] William Jackson to a friend, 12 January 1860; June 1860, KWM.
[20] Diary of William Allen, *Samuel Robertson*, 9 October 1844, Log 1040, WMLOD.

of the world, the possibility of making a fortune, the prospect of facing real danger. They found continued fulfillment in the whale chase, new challenges in deep-sea storms, and ongoing stimulation in foreign sights.

Silliman Ives explained that the sea's danger was a salve for his "wild and reckless" longings. Ambrose Bates admitted that he longed for "the excitement of a wild and dangerous life." "I do believe," he asserted, "that if I could just lower a boat and actack the ugliest whale that ever swam salt water I should not only conquer the said monster but I should come to the brig altogether changed. . . . Excitement Excitement Excitement what I crave in this my darkest hour."[21]

Whaling was as much hellish boredom as it was traumatic exhilaration, of course. Thus these mariners must have filtered out of their minds memories of fatigue, frustration, travail, and adversity. Some whalemen suggest that this was indeed the case. As ships neared the American shore after voyages of three to four years, the anticipation of freedom mollified social tensions and softened harsh memories. Sailors found new congeniality and discovered veins of humor and tolerance that had long been buried. Mate George Bowman felt the frustration of his voyage vanish even before his ship anchored. When a pilot came on board near Block Island on June 1, 1860, he and his shipmates smoothed over lingering differences in their excitement, and "all hardships and toils [were] soon forgotten." Whalemen also helped to nullify the adversity of a long voyage with physical means. Towards the end of a voyage, or even earlier if the vessel was full, sailors threw the tryworks into the sea, ejecting evidence of one of the most grueling aspects of shipboard labor. Sailors aboard the *Mercury* made the physical catharsis even more complete as they tossed aside articles of clothing and furniture. As they neared home, noted seaman Stephen Curtis, the crew threw over the ship's rails "mattresses, straw, hay, trousers whose looks betrayed the toil and anxiety of many a sleepless watch below, old shirts so completely covered with patches of all casts, shapes and sizes, that the original could scarce be discovered." All these, he said "followed each other in quick succession overboard, so that for a quarter of a mile astern, large bumps were seen floating on the surface of the waves, which a distant observer might have mistaken for the wonderful sea serpent."[22]

Once a man went to work on shore, he naturally compared his daily labors with shipboard life. It was then that the sea began to look, once again, like something magical. George Bowman got a job in a Massachusetts paint mill after a stint of whaling, where he got "along nicely" yet "still there is a desire

---

[21] Diary of Silliman Ives (under the pseudonym Murphy McGuire), 25 December 1868, Log 618, WMLOD; Diary of Ambrose Bates, *Isabella*, 12 August 1868; 10 September 1868, KWM.

[22] Diary of George Bowman, *Addison*, 1 June 1860, PPL; Diary of Stephen Curtis, Jr., *Mercury*, c. July 1844, NBFPL.

to go to sea again." John Spooner, who left the sea to work in a marble shop, found himself "growing restless again."[23] And Herman Melville, that well-known critic of clerical drudgery, described in his whaling novel how many men like himself, who spent their weekdays "pent up in lath and plaster – tied to counters, nailed to benches, clinched to desks" would, on a Sabbath afternoon, be "fixed in ocean reveries."[24]

Veteran Silliman Ives speculated on the attractions of deepwater sail to men recently returned to land. "Why is this thus?" he asked, when he found himself bound on another whaling voyage. "What in the name of the prophet possessed me to go again?" Later on he answered his own question and explained that there was "something irresistible in a sailors career":

> How many times after a long stop on shore have I found my way to some spot, where I could look out upon old ocean, and with the roar of the breakers in my ears and the ships as they went dancing over the waves visible to my eyes, there has come over me such a longing, such a desire to once more feel a deck under my feet, and to be tossed and tumbled about by old Neptune once again, as would scarcely be satisfied by all the inducements which a sojourn on terra firma could offer.[25]

Ives admitted that once he got to sea again the "glamour [was] soon banished" but that the trials and tribulations of seagoing were never vivid enough on shore to stave off the ocean sirens that called him back repeatedly.

Some mariners played over and over in their minds the conflict between sea and shore, seemingly internalizing the sense of themselves as social renegades. Thomas Morrison was one of several sailors who framed his seafaring career in terms of nearly pathological addiction. Bound home in early October 1873, he insisted that if he lived to get ashore "I wont go whaleing again so help my Bob by thunder." The following July he was back at sea again, and on the very first day out sounded a familiar promise: "If I do live to get home again I'll bet that I won't go whaling again." The following month he asserted that he really "will do different then I did this time and try to not to go whaleing again."[26]

"No inducement could persuade me to stay at home where ther is Everything that man could wish for," echoed a mate named Edwin Pulver, "but it seemed as if I must leav. I felt as if I was doing wrong. But go I must." Silliman Ives likewise hoped that in the future no temptation would "persuade me ever again to exile myself" at sea, and he prayed that "I may always continue in the same mind."[27]

And home with his wife Annie and young son Orren was, Ambrose Bates

[23] Diary of George Bowman (continuation of *Europa* log), 10 September 1871, PPL; Diary of John Spooner, *Mt. Wollaston*, November, 1863, NBFPL.

[24] Melville, *Moby Dick*, p. 12.   [25] Ives diary, 18 May 1868; 20 September 1868.

[26] Diary of Thomas Morrison, *Avola*, 3 October 1873; 16 July 1874; 16 August 1874, KWM.

[27] Diary of Edwin Pulver, *Columbus*, 9 December 1851, PPL; Ives diary, 6 August 1871.

The whaleman who incised a drawing on this tooth painstakingly celebrated a scene from domestic life. He claimed some symbolic space for
himself, however, in the ship that adorns the top of the tooth, and in the
schooner that sails in the distance, seen at left through the open window.

admitted, "all I could wish." Yet this man, more anguished than most, felt
driven to the sea again and again by a restlessness and a need for constant
diversion. At age thirty-five, aboard the bark *Millwood*, he wrote:

> Over and over again and I do not know but it always will be my lot to live
> in uneasiness In vain I have sought a place upon the globe that I might
> settle down in quiet contentment But alas none has yet lured me from the
> wild inclination for roaming And although I have been blessed with all the

heart could ask still as it seems almost against my will I find myself vo[l]inteerly flying from all I love on earth.[28]

A month later he translated his personal dilemma into verse:

> It was morning and I longed for evening
> It was evening and I longed for home
> I was at home and I longed for the sea
> Now whire shall I fly to amuse myself
> But alas [I] know of no place excepting
> I could combind Land and sea together.[29]

Ambrose Bates wished for nothing more than that he might combine the fulfillment of home with the satisfaction – the risks, the dangers, the aesthetic delights – of a sailing career. He and other veterans saw themselves as singularly divided, as uniquely torn apart, and as socially unhealthy. Their perception of themselves was to some extent shared. Domestic reformers ashore defined men who wandered for life as socially unwell in order to validate their efforts to bring them home and "cure" them.

But whalemen like Bates were not alone in their ambivalence. Men on the American mainland may not have gone sailing, but that does not mean that they did not seek oceans and shipmates. During the mid-nineteenth century many American males joined company and went west on overland trails. Together they adventured on the arid frontier, branding cattle, shooting buffalo, mining mountains, and facing down human adversaries. In the mid-1800s men also found comrades and tests of courage on battlefields in Mexico and in the American South. And every day they sought each other out closer to home, in Masonic lodges, in fire brigades, in militia companies, and in saloons.[30] Men on the mainland, like these men at sea, wanted a home yet yearned for adventure, hoped for security yet wanted diversion, and carried on the dual task of loving women and seeking worlds without them.

Within American culture, whaling was but one experience among many that provided men with an exclusive proving ground. The northeastern United States in the 1800s may have been distinctive in this regard, for it was a society caught up in enormous economic and social change that required of men and

---

[28] Diary of Ambrose Bates, *Milwood*, June 1867; August 1867, KWM. [29] Ibid.

[30] Carroll Smith Rosenberg, "Sex as Symbol in Victorian Purity: An Ethnohistorical Analysis of Jacksonian America," *American Journal of Sociology* 84 Supplement (1978): 212–47; Mark Carnes, *Secret Ritual and Manhood in Victorian America* (New Haven: Yale University Press, 1989); Elliot J. Gorn, *The Manly Art: Bare-Knuckle Prize Fighting in America* (Ithaca: Cornell University Press, 1986), pp. 142–3. Margaret Marsh, "Suburban Men and Masculine Domesticity, 1870–1915," *American Quarterly* 40 (June 1988), 165–86, sees a newly domestic man making his appearance in middle-class society in the early twentieth century. This man stands in contrast to the many men of the nineteenth century who spent their time apart from the family circle.

women significant personal readjustment. But it was certainly not the only society in which men practiced a life of paradox: celebrating freedom and responsibility, testing social boundaries and reaffirming them, pushing away from the families that nurtured them, and being forever pulled back.

Appendix I:
Description of Sailor Informants

Table I.1. Format of evidence

	Number	Percentage
Diaries (including logs)	194	92
Letters	16	8
Total	210	100

Table I.2. Branch of industry

	Number	Percentage
Whaling	172	82
Merchant shipping	38	18
Total	210	100

Table I.3. Place of residence before sailing

Town, region, or state	Number	Percentage
New Bedford region (including Cape Cod, Martha's Vineyard, and Nantucket)	50	33
Other coastal New England	67	45
Inland New England	9	6
Midatlantic region	24	16
Subtotal	150	100
Unknown	60	
Total	210	

Table I.4. Ages of sailors[a]

Age	Number	Percentage
12–14	3	6
15–18	8	16
19–22	16	33
23–26	13	27
27–30	4	8
30–40	5	10
Subtotal	49	
Unknown	104	
Total	153	

[a]At first departure, excluding mates and masters.

Table I.5. Ages of mates and masters[a]

Age	Number	Percentage
22–26	2	18
26–30	2	18
30–40	5	45
40+	2	18
Subtotal	11	99
Unknown	46	
Total	57	

[a]At first departure.

Table I.6. Position on shipboard[a]

Position	Number	Percentage
Boy	9	5
Greenhand	26	14
Foremost hands (unspecified)	29	15
Boatsteerers	13	7
Cooper/carpenter/shipkeeper	9	5
Cook/steward	6	3
Junior mates	13	7
Mate	24	13
Master	51	27
Passenger/observer	8	4
Subtotal	188	100
Unknown	22	
Total	210	

[a]At first departure.

Table I.7. Date of voyages at departure

	Number	Percentage
1825–1830	4	2
1831–1835	17	8
1836–1840	21	10
1841–1845	32	15
1846–1850	31	14
1851–1855	39	19
1856–1860	31	15
1861–1865	15	7
1866–1870	16	8
1871–1875	4	2
Total	210	100

Appendix II:
Age Composition of Whaling Crews

Table II.1. New London crews, 1818–1878

	Number	Average age
1818[a]		
Seamen of color	9	33
White seamen	62	25
Masters/mates	13	31
Total	84	
1828[b]		
Seamen of color	21	28
White seamen	104	25
Masters/mates	25	32
Total	150	
1838[b]		
Seamen of color	13	26
White seamen	198	23
Masters/mates	33	30
Total	244	
1848[a]		
Seamen of color	10	28
White seamen	189	23
Masters/mates	12	31
Total	211	
1858[b]		
Seamen of color	17	31
White seamen	138	23
Masters/mates	26	35
Total	181	
1868[b]		
Seamen of color	7	31
White seamen	106	26
Masters/mates	31	35
Total	144	
1878[a]		
Seamen of color	1	22
White seamen	82	26
Masters/mates	15	40
Total	99	
1888[a]		
Seamen of color	—	—
White seamen	19	27
Masters/mates	2	45
Total	21	

[a]Sample: all available. [b]Sample: every other white seaman. Seamen of color: all available.
Source: New London crew lists, Federal Archives Records Center, Waltham, MA.

Table II.2. New Bedford crews, 1818–1878

	Number	Average age
1818		
Seamen of color	66	28
White seamen	156	23
Masters/mates	13	31
Total	235	
1828		
Seamen of color	60	26
White seamen	483	22
Masters/mates	31	28
Total	574	
1838		
Seamen of color	23	26
White seamen	228	22
Masters/mates	13	28
Total	264	
1848		
Seamen of color	17	29
White seamen	263	22
Masters/mates	13	30
Total	293	
1858, 1868		
Data not available		
1878		
Seamen of color	188	24
White seamen	216	25
Masters/mates	19	35
Total	423	
1888		
Seamen of color	37	25
White seamen	65	26
Masters/mates	3	40
Total	105	

Source: 1848 only: Records of the New Bedford Port Society, New Bedford Free Public Library. All others: National Archives Microfilm, New Bedford Free Public Library. Sample: all records available or every other.

Appendix III:
Sailors' Place of Birth

Cautionary note: The low count of foreign-born sailors on the lists surveyed here probably does not reflect the full percentage of foreign-born men in whaling crews. Not only were citizens of other countries recruited during the course of the voyage but, because of a citizenship requirement on American vessels that went into effect in 1817, and that required crews to be two-thirds American, some sailors may have been quickly "Americanized" when they were shipped.

Table III.1. New London crew lists

	Number	Percentage
1828		
Coastal New England	151	64
Other U.S.	60	25
Europe	11	5
Other non–U.S.	3	1
Illegible/blank	12	5
Total	237	100
1838		
Coastal New England	221	47
Other U.S.	162	35
Europe	41	9
Other non–U.S.	21	4
Illegible/blank	23	5
Total	468	100
1848		
Coastal New England	113	51
Other U.S.	85	39
Europe	10	4
Other non–U.S.	8	4
Illegible/blank	4	2
Total	220	100
1858		
Coastal New England	131	39
Other U.S.	130	39
Europe	41	12
Other non–U.S.	29	9
Illegible/blank	2	1
Total	333	100
1868		
Coastal New England	77	30
Other U.S.	94	37
Europe	54	21
Other non–U.S.	22	9
Illegible/blank	9	3
Total	256	100
1878		
Coastal New England	55	38
Other U.S.	37	25
Europe	44	30
Other non–U.S.	7	5
Illegible/blank	3	2
Total	146	100

Source: Federal Archives Record Center, Waltham, MA. Sample: ten to fifteen crew lists per decade. Officers not distinguishable from other seamen. The voyages described in the years 1838 to 1868 were whaling voyages; the 1828 crew lists were

Table III.2. New Bedford crew lists

	Number	Percentage
1832		
Coastal New England	80	50
Inland New England	33	21
Other U.S.	31	19
Europe	8	5
Other non–U.S.	5	3
Illegible/blank	4	2
Total	161	100
1837		
Coastal New England	384	46
Inland New England	87	10
Other U.S.	265	32
Europe	33	4
Other non–U.S.	17	2
Illegible/blank	51	6
Total	837	100
1853		
Coastal New England	1061	30
Inland New England	236	6
Other U.S.	1245	35
Europe	64	2
Other non–U.S.	36	1
Illegible/blank	932	26
Total	3574	100
1858		
Coastal New England	911	37
Inland New England	161	6
Other U.S.	593	24
Europe	20	1
Other non–U.S.	6	—
Illegible/blank	784	32
Total	2475	100
1863		
Coastal New England	575	39
Inland New England	88	6
Other U.S.	411	28
Europe	17	1
Other non–U.S.	4	—
Illegible/blank	390	26
Total	1485	100
1868		
Coastal New England	646	40
Inland New England	23	1
Other U.S.	246	15
Europe	353	21
Other non–U.S.	226	14
Illegible/blank	154	9
Total	1648	100
1873		
Coastal New England	143	26
Inland New England	7	1
Other U.S.	76	14
Europe	123	22
Other non–U.S.	147	27
Illegible/blank	55	10
Total	551	100

Source: Records of the New Bedford Port Society, New Bedford Free Public Library. Sample: all lists from years described (every five years from the 1830s to the 1870s; 1840s not available).

Appendix IV:
National Composition of Ships' Crews

Table IV.1. Composition of ships' crews, 1 June–31 December 1850[a]

	Number	Percentage
Crews arriving directly from U.S.		
American citizens	246	79
Foreign citizens	65	21
Total	311	100
Crews departing directly for U.S.		
American citizens	483	79
Foreign citizens	128	21
Total	611	100
Crews inbound on American vessels from non–U.S. destinations		
American citizens	621	75
Foreign citizens	210	25
Total	831	100
Crews outbound on American vessels for non–U.S. destinations		
American citizens	580	69
Foreign citizens	256	31
Total	836	100

[a]As indicated on Talcahuano, Chile, consular returns, 1850.
Source: Talcahuano Consular Records, National Archives Record Service, Washington, DC. Sample: all available.

Table IV.2. Composition of ships' crews, 1 July–31 December 1850[a]

	Number	Percentage
Crews arriving directly from U.S. *on American vessels*		
American citizens	265	80
Foreign citizens	65	20
Total	330	100
Crews departing directly for U.S. Not available		
Crews inbound on U.S. vessels from *non–U.S. ports or destinations*		
American citizens	1711	63
Foreign citizens	1020	37
Total	2731	100
Crews outbound on U.S. vessels for *non–U.S. destinations*		
American citizens	1459	60
Foreign citizens	963	40
Total	2422	100

[a]As indicated on Lahaina, Maui, consular returns, 1850.

Appendix V: Shipmasters and New Bedford Ship Ownership

Table V.1.

	Number	Percentage
1851–1860		
Master is part owner	77	46
Master is not owner	90	54
Total	167	100
1861–1865		
Master is part owner	27	48
Master is not owner	29	52
Total	56	100
1866–1880		
Master is part owner	26	35
Master is not owner	49	65
Total	75	100

Source: *Ship Registers of New Bedford, Massachusetts*, Vol. II, 1850–1865; Vol. III, 1865–1939 (National Archives Project, n.d. [1930?]). Sample: random sample of vessels greater than 250 tons.

Appendix VI: United States
District Court Admiralty Records

Table VI.1. Literacy of mariner plaintiffs[a]

	Number	Percentage
1830–1840		
Signed	17	71
Marked	7	29
Total	24	100
1841–1850		
Signed	34	67
Marked	17	33
Subtotal	51	100
Not known	5	
1851–1860		
Signed	29	49
Marked	30	51
Subtotal	59	100
Not known	17	
1861–1870		
Signed	17	57
Marked	13	43
Subtotal	30	100
Not known	2	

[a]Literacy in this case is the ability to sign a name on a legal document. Record-keepers distinguished between signed names and "marks" in final record books.
Source: Final Record Books, United States District Court, Boston, Massachusetts, in the Federal Archives Record Center, Waltham, MA. Sample: all court records involving mariners at sea (with the exception of fishermen, steam sailors, and pilots) taken every four years from 1830 to 1870, with the exception of 1846, or as noted.

Table VI.2. Fines imposed on masters/mates
for criminal assault[a]

	Number	Percentage
Less than $15.00	5	23
$15.00 to $25.00	9	41
$26.00 to $50.00	4	18
$51.00 and over	4	18
Subtotal	22	100
Not guilty	10	
Case discontinued	5	
Total	33	

[a]1850–1870 only.
Source: Final Record Books, United States
District Court, Boston, Massachusetts, in the
Federal Archives Record Center, Waltham,
MA.

Table VI.3. Damages awarded to mariner plaintiffs for
assault or cruelty[a]

	Number	Percentage
1825–1840		
Awarded damages	19	76
Case dismissed/no award	6	24
Case dismissed by plaintiff[b]	0	
Total	25	100
1841–1850		
Awarded damages	16	80
Case dismissed/no award	1	5
Case dismissed by plaintiff[b]	3	15
Total	20	100
1851–1860		
Awarded damages	25	67
Case dismissed/no award	5	14
Case dismissed by plaintiff[b]	7	19
Total	37	100
1861–1870		
Insufficient data on litigation involving damages		

[a]Defendants were masters or mates. [b]May indicate out-of-
court settlement.
Source: Final Record Books, United States District Court,
Boston, Massachusetts, in the Federal Archives Record
Center, Waltham, MA. Sample: all records from 1825 to
1830; thereafter, all records every fourth year.

Table VI.4. Average awards for damages

	Average request	Average award	Award with court costs
1825–1840	$289	$29	$73
1841–1850	$515	$56	$113
1851–1860	$597	$57	—[a]
1861–1870	$750	$68	—[a]

[a]Not given.
Source: Final Record Books, United States District Court, Boston, Massachusetts, in the Federal Archives Record Center, Waltham, MA. Sample: all records from 1825 to 1830; thereafter, every fourth year.

Table VI.5. Suits for the recovery of wages (officers and seamen)

	Number	Percentage
1825–1840		
Full recovery (80–100% of request)	24	40
Partial recovery (40–79% of request)	14	24
Minimum recovery (1–39% of request)	9	15
Case dismissed by court/No recovery	11	19
Case dismissed by plaintiff[a]	1	2
Total	59	100
1841–1850		
Full recovery (80–100% of request)	22	73
Partial recovery (40–79% of request)	1	3
Minimum recovery (1–39% of request)	1	3
Case dismissed by court/No recovery	5	17
Case dismissed by plaintiff[a]	1	3
Total	30	99
1851–1860		
Full recovery (80–100% of request)	17	44
Partial recovery (40–79% of request)	5	13
Minimum recovery (1–39% of request)	4	10
Case dismissed by court/No recovery	6	15
Case dismissed by plaintiff[a]	7	18
Total	39	100
1861–1870		
Full recovery (80–100% of request)	3	12
Partial recovery (40–79% of request)	9	35
Minimum recovery (1–39% of request)	10	38
Case dismissed by court/No recovery	4	15
Total	26	100

[a]May indicate out-of-court settlement.
Source: Final Record Books, United States District Court, Boston, Massachusetts, in the Federal Archives Record Center, Waltham, MA. Sample: all records from 1825 to 1830; thereafter, every fourth year.

Index

Terms in *italics* are names of ships.

Warren, Joseph, 50

watch system, 60, 82, 121, 186

Watchman, 47

weapons: on shipboard, 11, 94–5, 97, 104, 108–9, 114, 122–3, 186, 190, 191, 192; on shore leave, 139

Weeden, Capt. Richard, 100–1, 112

Weir, Robert, 10, 56, 62, 63, 66–7, 70–1, 80, 84, 163, 188, 189; drawing by, 64, 66, 69, 70, 71, 135

Weld, Capt. Frederick A., 42–3

whale fishery. *See* Whaling

whale hunt: process of, 60–8; drawing of, 64, 66; as test of courage, 74; seen as primitive, 76; slowed down by sailors, 131

whale oil: 26; uses of, 17, 19; markets for, 19, 23, 25, 36; trying out, 68–72

whaleboats, 19, 60–1

whalebone, 26

whalemen. *See* Sailors

whales: Arctic bowhead, 20, 26, 37, 68, 72; California gray, 21; humpback, 21; right, 17, 20, 26, 66, 68, 72; sperm, 17, 20, 37, 64, 68, 72; depletion of, 26; scarcity of, 37

whaleships. *See* Whaling vessels

whaling: hardships of, 1–3, 4, 7, 73–5, 125; evolution of, 16–40; as different from home life, 21, 195–6; investments in, 22; attractions of, 46–57, 204–7; tradition of, in families, 51; in popular culture, 52–4; labor of, 56, 58–84; sailors' responses to, 59–84; tools of, 61; dangers of, 52–4, 68, 83–4, 132–3, 187, 196, 205; episodic nature of, 72; critics of, 75–80; social organization of, 81–4; compared to merchant marine, 133; as proving ground, 52–4, 207–8

whaling fleet, 6, 36, 37–9

whaling grounds, 21, map of, 38

whaling industry: peak of, 6, 16–23, 36; beginnings of, 16–17; expansion into Arctic Ocean, 21; in New England economy, 21; operation of, 23, 33–6; mid-nineteenth-century changes in, 24–7; shifts to Pacific, 26; profits of,

36, 47; decline of, 36; investors in, 39; and labor costs, 92; and Congress, 93–6

whaling products, 24–6

whaling vessels: design of, 27–32, 33; compared to factories, 31; diagram of, 32; cost of, 33; photo of, 34; as chandleries, 35; role of, in blockade of Southern ports, 37; as female figures, 174–9; design of, 194

White Act, 93

Whitfield, Dan, 51, 55, 151–2

Wilbur, Sanford, 190

Wilcox, Henry, 152

Willcox, Holden, 181, 183

William Gifford, 113

William Wirt, 48

Williams, Eliza, 168

Williams, Lewis, 12, 48, 127, 164, 168, 189

Wilson, Mrs., 164

Wilson, William, 112, 137, 159

Winchell, Capt., 132

Winegar, Capt. Samuel, 85, 105

wives of captains: 31, 98, 99–100, 124–5; and sailors, 163–8; favoritism by, 164; diaries of, 166; pregnancy of, 166; and separation, 201–3

wives of sailors: 198; infidelity of, 180–1; and separation, 202

women's work. *See* Domestic work

women: 162–94; and attitudes toward sailors, 76–7; as humanitarian influence, 87, 96; and influence on shipboard behavior, 98–100; native, and captains, 113–14; native, and sailors, 133, 139–40, 148–53, 154, 157–9, 164, 170, 181–4, 187–8; beauty of, and race prejudice, 157–9; as sailors, 162; whaling an escape from, 164; power of, 168, 203; rituals concerning, 173–4; depictions of, 174–8; infidelity of, 180–1; at home, sketch of, 182

work ethic: of sailors, compared to indigenous peoples, 159–60

Zone, 117